Growth on a Finite Planet

by Lindsey Grant

SEVEN LOCKS PRESS
SANTA ANA, CALIFORNIA
MINNEAPOLIS, MINNESOTA
WASHINGTON, D.C.

Library of Congress Cataloging–in–Publication Data
Grant, Lindsey.
    Juggernaut : growth on a finite planet / by Lindsey Grant.
      p. cm.
    Includes bibliographical references and index.
    ISBN 0-929765-49-4 (hardcover). — ISBN 0-929765-51-6 (pbk.)
    1. Population policy. 2. Population research. 3. Overpopulation.
I. Title.
HB849.4.G73   1996
363.9—dc20   96-35264
                                   CIP

Manufactured in the United States of America

Seven Locks Press
P.O. Box 25689
Santa Ana, CA 92799
(800) 354-5348

Book design and typography:
Joel Friedlander Publishing Services

This book is dedicated to Lee Bouvier,
old comrade in arms.

ABOUT THE AUTHOR

LINDSEY GRANT writes on population and public policy. A retired Foreign Service Officer, he was a China specialist and served as Director of the Office of Asian Communist Affairs, National Security Council staff member, and Department of State Policy Planning Staff member.

As Deputy Assistant Secretary of State for Environment and Population Affairs, he was Department of State coordinator for the *Global 2000 Report to the President*, Chairman of the Interagency Committee on International Environmental Affairs, U.S. delegate to (and Vice Chairman of) the OECD Environment Committee and U.S. member of the UN ECE Committee of Experts on the Environment.

His books include *Foresight and National Decisions: the Horseman and the Bureaucrat, Elephants in the Volkswagen* (a study of optimum U.S. population) and *How Many Americans?*

CONTENTS

The Author vi
Acknowledgments ix
Introduction 1

PART I

THE TECTONIC FORCES 5

Chapter 1. Population. The Unique Century 7
Chapter 2. The Prospects for Food 17
Chapter 3. The Energy Transition 45
Chapter 4. Doomsday Scenarios: Energy, Pollution and Climate 56
Chapter 5. Technology: Deus ex Machina? 78
Chapter 6. A Conserving State of Mind 92

PART II

INTERLOCKED CONSEQUENCES: A DARKENING CENTURY 101

Chapter 7. Diverging Futures 103
Chapter 8. One World, Like It or Not 132

PART III

THE UNITED STATES: ISSUES, TEMPORIZATIONS, SOLUTIONS 155

Chapter 9. U.S. Population Growth: An Accidental Future 157
Chapter 10. Living with the Land 163
Chapter 11. The Collapsing Society 173
Chapter 12. Population and Policy 184
Chapter 13. The Population Solution 203
Chapter 14. The War of the Paradigms 217
Chapter 15. Of Tigers, Ants and People 226
Chapter 16. Multiple Agendas and the Population Taboo 237
Chapter 17. The Failure of Leadership 251
Chapter 18. Creating a Consensus 264
Conclusion. A Somber Optimism 280

Notes on Sources 284
Geographical Conventions 285
Notes 287
Glossary of Acronyms 302
Index 304

FIGURES

1-1. World Population History 7
1-2. World Population Projections 9
1-3. Century of Change 14
1-4. Chemicals: Health Hazard Assessments 15
1-5. Population by Region 16
2-1. U.S. Corn Yield, 1866–1993 24
2-2. World Grain Use, 1960–1992 37
2-3. World Fish Catch, 1938–1993 39
3-1. Petroleum Futures, 1992 46
3-2. Gas Futures, 1992 47
5-1. Auto Inefficiency 81
7-1. Comparative Population Pyramids 104
7-2. Constant Fertility Projections 106
7-3. Africa's Best Hope 109
7-4. AIDS and Population Growth 111
8-1. Third World Working-Age Population 133
9-1. U.S. Population, 1790–1990 157
9-2. U.S. Population, 2000–2100 158
9-3. U.S. Population Projections,
    Constant Fertility, Mortality, Migration 160
11-1. Young and Idle 174
11-2. U.S. Wages and Productivity 176
12-1. Alien Work Force Entrants vs. Job Growth 185
12-2. U.S. Population 65 and Over 196
13-1. The Two-Child Family 204
13-2. Fertility of U.S. Women 208
18-1. A Revised View of the Economy 275

ACKNOWLEDGMENTS

I wish to thank the S. H. Cowell and Weeden Foundations and a most generous but resolutely anonymous donor for their support of the costs of writing, editing and promoting this book. The Center for Immigration Studies and Negative Population Growth administered those grants. I am indebted to one of the great figures in the population movement, Garrett Hardin, for his comments on the manuscript and to Max Thelen, Jr., and Ernest Callenbach for reviewing the draft. Among many people in and out of government who were helpful, let me name only a few. Dr. Tom M. L. Wigley of the National Center for Atmospheric Research, with the able help of Ms. Lisa Butler, gave me invaluable help in obtaining current documentation on the climate issue. Dr. Ronald Charpentier of the U.S. Geological Survey was most helpful on fossil energy data, and Drs. Edward Overton and Noel Gollehon of the Department of Agriculture on crop and water data. Dr. David Pimentel was helpful as always, and I am grateful to Dr. Leon Bouvier for his friendship and advice and for reading and offering comments on the manuscript. None of those people, of course, are responsible for what I have written. Let me add a note of appreciation to Publisher James Riordan and his colleagues at Seven Locks Press, a thoroughly congenial group to work with.

Some sections of the book were drawn from monographs I wrote for Negative Population Growth, Inc's NPG FORUM series. A special salute is due to Donald Mann, President of NPG. For a quarter century, he has steadfastly maintained that the populations of the world and the United States are already too large. In the face of timidity even among his allies, he has insisted on calling for "negative population growth." Public Relations experts shudder at a title starting with "negative," but he is right, and he has helped to widen the range of acceptable debate about population.

Lindsey Grant
Santa Fe, 1996

INTRODUCTION

An irreverent thought crossed my mind as I was looking recently at a survey of publications about world issues: "Do I really want to add another book to all this?" I convinced myself that I did, but only for one reason: very few of those books touch upon population as a source of the issues they address, and fewer still identify population policy as a part of the solution. Except for a hardy handful of population writers, the authors do not make the critical interconnections among population, consumption, environmental problems, unemployment, social issues and political crises. They identify every problem except the central one; they preach every solution except the most accessible one: the conscious management of human fertility.

If there were one message I could imprint upon the American mind, it would be this:

Perpetual physical growth is impossible on a finite planet.

The statement is mathematically self-evident, but it is simply ignored. Once one accepts it, the question inescapably becomes, not Should human population growth stop? but When? How should it stop? By conscious policies to limit fertility or by rising mortality returning again to the levels of the bad old days? The distinguishing characteristic of this century is that growth has reached the point at which this choice has become an immediate issue.

This book is meant to show the connections between population change and the major issues that face the world and to suggest how we in the United States can match our demography to our aspirations, and our aspirations to demographic realities.

Keep an image before your mind: a child crying because it is hungry. That did not have to happen. The human tribe does not need to go that way. Yet, right now, perhaps 800 million people literally go hungry.

There can be a vision of a world in which there is enough for all and the promise of dignity. The vision rests—perhaps I should use the past tense: rested—upon the idea that with technology tools become our slaves, and civilizations do not have to be built upon the back of human toil. That vision is growing dim because population growth is eating up the potential surplus that technology has helped us to discover or create. In more limited contexts, humans have gone through that process before and then moved on to successive New Worlds. Now, there is no place to run.

There is something new on Earth: the scale of human activities. Part of it reflects rising consumption in the industrialized countries, part of it an unprecedented spurt of population growth in the "third world" (see "Geographical Conventions")—a result largely of the rapid application of disease control and agricultural technologies developed earlier in the West.

Any humane observer would sympathize with the desire to bring mortality down and to keep children from dying. Nevertheless, the demographic balance was profoundly disturbed when mortality was lowered while nothing was done about fertility. World population has doubled in less than 40 years, and it will probably double again—if the Earth can sustain it—even if the developing world succeeds in reducing fertility. Third world leaders trying to slow the surge of population growth must bitterly wish that they had started earlier. The ironic point is that demographers could have told policy makers what the results of their policies would be. But nobody asked. Now we must strive to ameliorate rather than avert the consequences of that growth.

In a recent book, I invoked the metaphor of the Juggernaut:

> In South India, at a certain festival, it was customary to roll out a giant ceremonial chariot in honor of Lord Vishnu. The chariot was wheeled, but no provision was made for steering it. Pulled and pushed by devotees in religious ecstasy, it crushed everything in its path: people, animals, buildings. It was called the Juggernaut.
>
> Population growth has something of that same terrible appearance of inevitability. However, like the Juggernaut, it is driven—not by fate—but by humans who are unconcerned about the consequences of its progress. If we can bring those driving the population Juggernaut to consider the consequences of what they are doing, perhaps we can turn it from its present destructive course.[1]

Humankind is fast descending into a time of troubles, and the troubles are largely of our own making. The human tribe, or parts of it, has faced disasters before, but this is not a natural disaster. We can stop the Juggernaut if we will recognize the sources of our problems and mobilize our fractured energies.

Unlike the Indian Juggernaut, population growth is invisible to most people. The numbers are visible enough. Try a visit to Calcutta or New York. But the growth is almost imperceptible. We try to adjust to its gradual effects rather than confront it. History is a slow-moving panorama, and each human peers at it through the narrow lens of one lifetime. As a species, we do not handle long-term problems very well. Our early ancestors survived by their ability to escape a predator or catch a fast-moving prey, not by contemplating environmental change. It is the human propensity to accommodate to population change rather than address it that makes this particular Juggernaut so dangerous.

Population growth is leading us to a world that we do not want. It is the most fundamental of the engines of change, and the most ignored. The poor nations face sheer hunger and the destruction of their resources. The "emerging nations," most of them in Asia, are in varying degrees escaping those horrors to face the problems of industrialization. The old "rich" countries confront joblessness, failing social structures, growing disparities between the rich and the poor, ethnic conflict, the loss of a shared vision, environmental degradation and the huge reality that they are changing the climate we all live in. Bringing population growth under control will not necessarily solve those problems, but it is the condition precedent—a necessary condition for their solution.

The economists say "we must grow out of the problems." We cannot grow out of them, not at this stage of human history.

As William Catton put it, we were living in the "Age of Exuberance."[2] The end of growth is a very tough change to face. U.S. political leaders, almost without exception, would rather ignore it. Whether we face it or not, the Juggernaut will stop. If not by choice, then by breakdown.

In my mind's eye, this book is an extended essay, written for the college undergraduate, offering a connected vision of a complex and fast-moving world. I hope that it will be of interest also to any reader undertaking to sort out the welter of conflicting evidence as to where the world is heading at the close of a most eventful century.

I have assembled the specialists' best judgments as to what is happening in different fields and then made the connections—what is happening in different places and sectors and how those changes interact. Some of what I argue cannot be rigorously proved. Causation is very hard to prove in complex systems, and most human decisions are based not on absolute proof but on common sense and the preponderance of evidence. Let me offer a variation of Occam's Razor: trust your common sense unless the weight of scientific evidence is overwhelmingly counterintuitive.

Don't trust statistics. You will see many statistics in this book. If you are suspicious of them, rest assured that I am, too, and from personal experience. Spurious precision is a weakness of statisticians. A figure of 73.156% means very little if the margin of error is 10%. Statistics vary widely in how closely they track the real world. I believe, for example, that live births are reported with rather high accuracy in the United States, but deaths may be somewhat less accurately reported, particularly when there is a Social Security pension involved, or foul play, or an identity—a name and Social Security number—that can be sold to smugglers of immigrants. I suspect that we have only an imprecise idea of how many people there are in American cities. In some other countries, vital statistics may be no more than guesses. Projections are even worse; they are no better than the assumptions on which they are based.

Nevertheless, we must quantify to understand what is happening. Even shaky data may show in broad terms which way we are going. I use the most widely accepted official data (see "Note on Sources"), if only to avoid contentious side issues. The official data and projections are ominous enough.

In any event, *trend* is more important than precision. The absolute numbers are less significant than the direction they are moving. Are things getting better or worse?

The fundamental human needs are not much discussed by futurists: food, water, shelter, work and dignity, a livable environment, and some confidence that they will last. I will focus on those issues and how they are being shaped by population change.

I propose to

- start with global forces and their interactions,
- describe their varying impact in the poorest countries, the emerging industrial countries and the industrial world,
- show what those changes mean for the pursuit of U.S. national policies (e.g., environment, jobs, trade, welfare, health), and
- suggest how to stop the Juggernaut.

It is an interdependent world. The atmosphere. The biosphere. Climate and world sea levels. Food supplies. Trade and employment. I will start with the world, but my purpose is to offer ideas as to how Americans might overcome the problems that confront us. On the principle "think globally; act locally," we must face the question, What can we do about it? here in the United States. Most of the proposals for action are thus in Part III.

Prediction is guesswork. The world is too complex for anybody to see the future. Still, we need to look at the vectors—the directions in which the world is presently heading, and the speed. How do human population behavior and our pursuit of economic, environmental, social and political policies intersect with each other? Where are they leading us? Do we want to go there?

PART I

The
Tectonic
Forces

I BORROW THE WORD FROM THE GEOLOGISTS. "Tectonics"—from the Greek word for a builder or carpenter—are the basic forces shaping the major landforms of Earth.

The issues that others consider basic, I find derivative. Wars. Insurrections. Hunger. The breakdown of social standards and political processes. The decline of education. The eruption of racism and ethnic conflict. Even joblessness. These can seem to be of transcendent importance, but I would argue that most of those problems have been generated or immensely exacerbated in the past half-century by four titanic and interconnected forces:

- accelerated growth of population and consumption,
- growth in food demand and output,
- technological change, and
- extraordinary use of fossil fuels.

Their interaction has generated deep instabilities in world environmental and socioeconomic systems.

Academic commentators and governments have recognized these forces only dimly. The synergies—the ways in which the forces drive each other—have been largely ignored, but those underlying forces and interactions will shape nations' fates over the next several generations.

The physical and social systems that make up the Earth and human society are immensely complex. When we change them, we don't know what we are doing. I don't, and neither do business or national political leaders. The difference between us is that they plow ahead into the darkness. My plea is for caution. There is something called the Rule of the Prudent: act on the more cautious estimate of a problem. You will be safe if you were right and pleasantly surprised if you were wrong. If you chance it on an incautious estimate, you may be ahead if the gamble works, but face a disaster if it doesn't.

CHAPTER 1

Population.
The Unique Century

THE FIRST FORCE IS POPULATION GROWTH, since it is people who consume, people who pollute, and humans who, more than any other species, are changing the Earth's ecosystem right now.

WHERE WE ARE

The present is unique. Human populations have grown about three times as much in this century as in all preceding time. The curve is something like this:

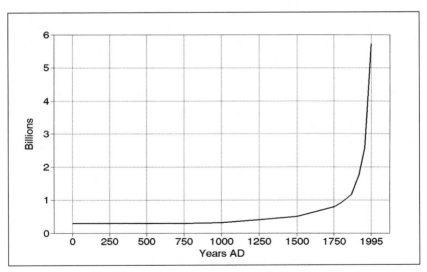

Fig. 1-1. World Population History
(UN June 6, 1994)

The present spurt of growth has been achieved with an even faster growth in the exploitation of energy, chemicals and the "renewable resources" of Earth, all of them supporting unprecedented consumption in the industrial world. *Never in human history have population and consumption levels changed so fast or reached current levels.* That growth is the source of the issues discussed in this book.

WHERE WE'RE HEADING

Any material growth in a finite world eventually becomes absurd. The distinguishing—and usually overlooked—characteristic of the present is how close we are coming to the edge of the absurd. I cited three such absurdities in an earlier book:[3]

- At the rate consumption was growing just before the 1973 oil crisis, all the world's oil would have been exhausted in 342 years, even if the entire planet were made of oil.

- Increase the rate of consumption of anything by just 2% per year (very roughly, the world population growth rate) and a "million-year supply" at current rates becomes just 501 years' supply.

- At that population growth rate, human density would reach one person per square meter of ice-free land in about 600 years.

To come closer to the present: the United Nations' (UN) world "constant fertility" projection for 2150 (figure 1–2) shows what would happen if fertility in the various areas of the world should stay at the 1990 levels. That figure translates to about 0.019 hectare (roughly 45 by 45 feet) of land per person, including deserts, mountains, glaciers, forests, plains, farms, cities and highways.

There has recently been a wave of false optimism that somehow the problem is on the way to solution. Much of it originated at the UN International Conference on Population and Development (ICPD) held in September 1994 in Cairo. The conference concluded (without elaboration or justification) that its Programme "would result in world population growth during this period (1995–2015) and beyond at levels close to" the UN 1992 low projection and that it would stabilize world population in the next century. The Programme is more optimistic than its technical advisers for reasons I will describe in chapter 17.[4]

Let us look at the UN projections in figure 1–2.

That "low projection" in the graph would indeed lead to a world population peak of only 7.8 billion in 2050 and a subsequent turnaround. The problem is that it is a remarkably optimistic scenario. Take Africa, the most desperate continent, as an example: it would require that average fertility

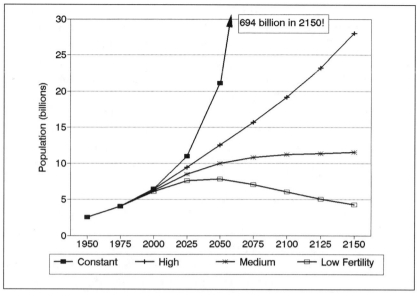

Fig. 1-2. World Population Projections
(UN 1992 Projections)

of women (total fertility rate or TFR) decline from 5.9 children now (and 6.4 in sub-Saharan Africa) to 2.31 in 2015–2020 and 1.6 in 45 years—a 70% reduction in a region that has achieved only a 10% reduction since 1950. For India, the required decline is from 3.75 now to 1.5 before 2030. The low projection is a tough and unlikely scenario even with massive efforts on all feasible measures to reduce fertility. Nevertheless, I will use that low projection as a stalking horse in coming chapters because it is a vision of what lies just within the bounds of the possible, with good luck and a worldwide commitment to doing something about population growth.

A difference of 10% in average world fertility—between 2.0 and 2.2 children per woman—is the difference between eventual stabilization and continued growth toward the mathematically absurd. A difference of one child (between two and three, for instance) is a difference of billions of people in two generations—and fertility surveys indicate that most women in poor countries presently want more than three children.

Let us turn to that astonishing constant fertility projection. What it dramatizes is that the so-called third world is less than halfway to manageable fertility levels. The war is far from won. Excluding China (which has had remarkable success in reducing fertility), overall fertility in those countries has declined less than one-third since the 1950s to 4.2. It must still be cut in half again, to below 2.1, if population is *ever* to stabilize. (That is, barring a rise in mortality, which is quite likely.) That's a crisis.

All other long-term world population projections (UN, World Bank, Census Bureau) are optimistic. They all assume continuing declines in third world fertility. They simply posit different timetables. There is some reason for hope. Fertility has been falling in most of the third world. As we have seen above, however, the task ahead is enormous. Arithmetically, growth is faster right now than it has ever been, about 90 million each year.

The UN demographers remark that the constant fertility projection is unrealistic. Of course it is; it's absurd. Famine and disease will arrest that curve before it goes very far. For that matter, the "high" and "medium" projections (figure 1-2) may be nearly as unrealistic. There is no assurance that a world of deteriorating environment and resources can support 28 or 12 billion people, particularly if people in the newly emerging countries insist (quite understandably) on trying to live like we do in the industrial countries. In coming chapters, I will try to show that even the present 6 billion, consuming and polluting as we do, is unsustainable.

There is an endearing optimism about world population projections. They assume life expectancy will continue to rise toward some biological limit—i.e., a modern, hygienic world, without hunger, where childhood diseases are under control and the mortality of the old continues to decline. The 1994 UN median projection for Africa has life expectancy rising from 53 years now to 73 years by 2045–2050, although per capita food intake in Africa has been declining for decades, and Africa now faces massive mortality from AIDS (chapter 7). Statistics reflect the beliefs and assumptions of the culture that creates them, and bad news is not popular.

Having made those assumptions, demographers project various levels of fertility for different countries, usually converging at some time toward a TFR of about 2.04 to 2.1. That is, roughly, an average of two children per family (plus a very small allowance for mortality and for the biological reality that fewer girls are born than boys). This is considered "replacement level fertility," the level that would produce a stable population with the low mortality the demographers have assumed.

In other words, these magic mortality and fertility projections assume a wonderful future, irrespective of population growth or its consequences.

In fact, the "2.1" myth is just that: a myth. An average total fertility rate of 2.1 is not foreordained, nor would it necessarily lead to population stabilization. In the real world, fertility has always fluctuated. In earlier eras, replacement level fertility may have been five or more children to keep up with mortality. If we push our support system—arable land, the environment—too far, or if we cannot control disease, mortality will rise again. There is no way of knowing what combination of fertility and mortality will occur in different countries or what population it will lead to.

The statistical conventions about mortality and fertility may be inescapable. One can imagine the hornet's nest the statisticians would poke into if they ventured assumptions as to the effects of population growth on mortality, so they make none even while they recognize the frailty of the conventions they are using. This is a superb example of the ways in which we are constrained by conventional expectations from exploring the real issues confronting us. It suggests, however, a certain caution in reading the projections. If they seem to show that the Earth can support a given population, it is because the proof is circular.

Future populations will be determined either by human volition or by natural causes—by limiting fertility or facing rising mortality. The constant fertility curve makes the crucial point that, one way or the other, human population growth must stop, and very soon. The limits I will describe in later chapters suggest that it will stop within the next century. The issue is how much damage we will do to the Earth as a habitat for humans and other beings before it stops.

The central policy question before the human race should be this: How do we continue the progress made so far and get fertility down before rising mortality takes over again and determines our demographic future?

DIFFERING FUTURES

There are vast differences between what is happening in different parts of the world. They are driven mostly by different levels of fertility, and we cannot assume that fertility levels will automatically converge toward that magic 2.1.

Europe is presently heading toward population decline, with long life expectancy. The emerging industrial countries face very different demographic futures. Some are becoming demographically indistinguishable from the older industrial countries, while others are splitting into rich minorities in seas of poverty. The poorest countries are heading into the limits set by rising mortality.

Those trends may change because of their interaction. The wild card is migration. Like hydrostatic pressures, fertility differentials drive migration, and immigration has become the major demographic driving force in the United States and West Europe and a growing element in the emerging industrial countries.

The different patterns of fertility are closely related to the industrialization process. It has offered jobs, education and opportunity to women that they did not have in traditional societies. Exactly what causes fertility decline is hotly debated but probably not provable. However, deliberate policy can lower fertility by making contraceptives and abortion available, publicizing the benefits of smaller families, and in some countries (such as Singapore), offering inducements to keep family size small. Some countries have managed substantial reductions in fertility without industrialization, or before it took hold—China, Indonesia, Cuba, North Korea, Bangladesh, Botswana, Tunisia, Jordan, Nicaragua, Zimbabwe, Morocco, Kenya, to name a few.

On the other hand, fertility in industrial countries has tended to move with attitudinal changes and women's sense of what kind of world their children would be born into. (Fertility in Russia and Eastern Europe has fallen very low in the present difficult times.) Industrial countries have had very little experience with deliberate efforts to lower fertility except in Japan, where a sharp decrease took place, starting just after World War II, quietly supported by the government and achieved largely through widespread abortion. Governmental attempts to raise fertility in industrialized countries have been of dubious effect, presumably because they have not succeeded in enlisting the people in the effort.

URBANIZATION

The industrialized world is 75% urbanized, and its cities are likely to grow only 10% in numbers by 2025 (UN 1994 projections). That is a manageable prospect for cities with their support infrastructure in place. But be warned: most of the projections assume little or no migration from the third world, even though cities like Los Angeles are growing largely because of such migration.

The third world is urbanizing under very different circumstances. Third world population growth accelerated around 1950. The peasants soon ran out of land and in desperation moved to the cities in rising numbers. The process is continuing. The urban population of the third world has increased six-fold since 1950, adding 1.4 billion people. It is expected to double again by 2025, adding another 2.3 billion people in even less time. These figures are almost incomprehensible. The equivalent of two Chinas, or 40% of the world's present population, will be piled upon cities that are already in or approaching breakdown. The highest rates of urban growth are expected in the poorest countries that are least prepared to accommodate such growth.

The continuing influx hopelessly overloads the cities' infrastructure. Cities that had been able to provide water, sewage systems, electricity, education, justice and fire and police protection to most of their populations are overpowered as shantytowns grow in every vacant corner. As factories spring up to exploit the cheap labor, air pollution soars. Differential wealth seems to be a condition of modern civilization, and the number of automobiles and trucks has risen spectacularly in third world cities despite the poverty, producing transportation gridlock, more pollution, and a grim environment in which the poor must live and work. Mexico City, already close to unbearable, has had to close a refinery and ban automobiles on alternate days.

Crowding in cities without piped water and sewers is a prescription for the spread of epidemics, and they have been spreading. The diseases of crowding and filth have been reappearing: cholera, dysentery, typhus, typhoid, tuberculosis, even the plague—diseases we had thought were scourges of the past. A recent UN report, surveying the projections, warns that the problem will get worse. Lack of sanitation, which receives less attention than air pollution, "usually has a far greater toll on . . . health."[5] The future of third world cities bears all the marks of a catastrophe, and the effects will not stop there (chapter 8).

There is perhaps one glimmer of benefit: fertility is usually lower in the cities, perhaps because children are a more apparent burden there than in the countryside. Urbanization thus slows population growth, but this relatively benign population check is much less important than another prospect: pestilence and rising mortality will be grim reminders that if we cannot manage our fertility, nature will manage our mortality.

THE UPSTART SPECIES

Humans are one of the newest species on Earth, and we are changing the Earth more rapidly and fundamentally than any other species.

Why does it matter?

Most fundamentally, the human tribe is on the way to diverting most of the Earth's resources to our own use. We are inadvertently encouraging organisms we don't like and wiping out those we like or should like. We are changing the ecological systems that support us, without understanding quite what we are doing or what the consequences will be. The fact that we can change the Earth is no guarantee that we can manage it. We have assumed the management—and we are doing it very badly, not just from the standpoint of other living things, but from our own narrow interests.

To visualize recent experience, consider figure 1-3 below.

Compare the population curve here with that in figure 1-1. This time, it looks comparatively modest because it covers just one century, and population growth is less dramatic than the remarkable growth in energy,

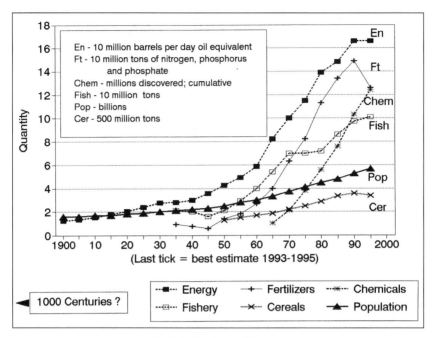

Fig. 1-3. Century of Change

fertilizers, fisheries and chemicals. Population growth drives those other curves as growing populations try to sustain rising consumption levels.

When I first drew this graph a decade ago, the question was whether those dramatic growth curves could continue at such rates. Apparently, we are getting the answer earlier than I expected. The growth in energy use and fertilizer may resume as the emerging nations' demands kick in (chapter 7), but except for aquaculture, the fishery catch is declining.

The two critical lines are population and cereals. If food production cannot keep up with population growth, that growth slows or stops, and all our technical toys, our electronic wonders, even the advances in medicine, become irrelevant. Cereal production did keep up for most of the century, but it has fallen behind in the past decade. We don't yet know whether the shift is permanent, but it cannot simply be explained by a run of bad weather.

Man–made chemicals are a particularly troublesome case of our bad stewardship. Far more chemicals have been introduced in the past three decades than in all preceding history. Some 60,000 are in common use. The National Academy of Sciences (NAS) in 1984 sponsored a sample survey to determine how many chemicals had been studied for their toxic effects on people or animals. Here is what they found.

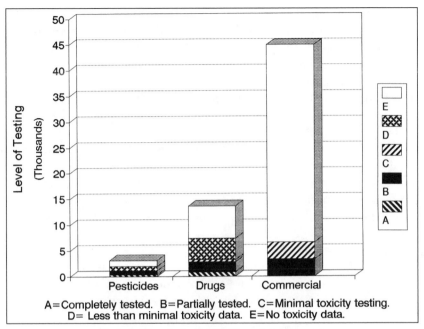

A=Completely tested. B=Partially tested. C=Minimal toxicity testing.
D= Less than minimal toxicity data. E=No toxicity data.

Fig. 1-4. Chemicals: Health Hazard Assessments
(NAS/National Research Council 1984)

("Drugs" includes cosmetics and food additives.)

The great majority of chemicals we are ingesting or dumping into the biosphere have not been tested for direct toxicity to humans. The study did not investigate the environmental and health effects downstream as the chemicals interact with each other and with natural processes and change into other compounds. The organic chlorine compounds, for example, are particularly likely to undergo such changes. In a complex system, we simply cannot track all the implications. The National Academy of Sciences and the Office of Technology Assessment (OTA) have warned of the unknown dangers from chemical mutagens—which might not surface for "several generations"—and neurological toxicants.[6]

The frightening conclusion is that

we don't know what we are doing.

Subsequent chapters will describe similar uncertainties as to the consequences of most other economic activities, particularly in agriculture and energy production. Anybody who has listened to the raging scientific debates as to just how much damage is being done must necessarily come away with at least one conclusion: we aren't sure. And that *uncertainty itself should warn us of the need for caution.*

How big is big? Even granted that human activity has multiplied, how does this affect the Earth? We really don't know, because there are few baselines. When our ancestors should have been taking measurements, it did not seem like a potential problem. It is clear that the current levels of activity are affecting the Earth. I will relate some of the ways: rising atmospheric carbon dioxide, methane and other pollutants and greenhouse gases; land loss and degradation; desertification; salinization; forest declines; and declining water per capita in dry regions. Many other indicia are at least as dramatic, but the evidence is partial, contentious, anecdotal or local: wastes, air quality, the disappearance of animal species at a rate unmatched since the Cretaceous extinctions, the decline or disappearance of fish stocks and marine mammals, the mysterious rise in human health problems such as asthma, Alzheimer's disease, and perhaps cancer.

For readers' convenience, since population comparisons will be appearing through most of this book, it may be useful to summarize the recent population history of the Earth's major regions and countries and the fertility levels that are presently driving further growth (figure 1-5).

POPULATION BY REGION
(UN 1994)

	1950	1995	Fertility
	(Millions)		(TFR 1990-95)
World	2520	5726	3.10
Africa	224	728	5.80
(Nigeria)	33	112	6.45
(Ethiopia)	19	55	7.00
(Zaire)	12	44	6.70
(South Africa)	14	41	4.09
Asia	1402	3458	3.03
(China)	555	1221	1.95
(India)	358	936	3.75
(Indonesia)	80	198	2.90
(Pakistan)	40	141	6.17
(Japan)	84	125	1.50
(Bangladesh)	42	120	4.35
Europe	549	727	1.58
(Russia)	103	147	1.53
(Germany)	68	82	1.30
(United Kingdom)	51	58	1.81
(France)	42	58	1.74
(Italy)	47	57	1.27
(Ukraine)	37	51	1.64
Latin America	166	482	3.09
(Brazil)	53	162	2.88
(Mexico)	28	94	3.21
(Columbia)	12	35	2.67
(Argentina)	17	35	2.77
North America	166	293	2.06
(United States)	152	264	2.08
(Canada)	14	29	1.93
Oceania	13	29	2.51
(Australia)	8	22	1.87

Fig. 1-5. Population by Region

Chapter 2

The Prospects for Food

IN THE COLLISION BETWEEN expanding human pop–
ulations and a finite Earth, food will probably be the limiting factor that
brings a halt to population growth. The limits for food in turn are set by
the availability of arable land, water and energy and by human ingenu–
ity in raising yields. (In this book, I will use cereals as a surrogate for all
vegetable foods since they are the major component and the easiest to
track statistically.)

Like Sisyphus pushing his stone, we are trudging uphill. After decades
of successes, the world faces more ominous threats than before in trying
to provide enough food for more people. Grain prices have risen sharply,
and the worldwide stock carryover for 1996 is down dangerously to 49
days' supply, the lowest since full records became available. As a result, a
failure of the monsoon rains in China or India or a bad year such as the
United States experienced in 1988 could create a panic in the world grain
market and lead to massive starvation (chapters 7 and 8). Population
grows; the Earth does not.

THE LAND

The basis for food production is land—space for plants to grow with
enough sunlight, warmth, water and nutrients to support life and a soil
matrix to enable them to use those inputs.

World population has grown 35% since 1970. The world's farmed
acreage has not much changed, and it is deteriorating as good lands are
lost to other uses, to erosion or desertification, and poorer lands are put
to the plow. The amount of land *per capita* declines with population
growth—an obvious statement but one that is regularly forgotten. To turn
the statistic around: arable land per capita has declined 35% since 1970,
and it is poorer land. It faces another 40% decline in the next two

generations from population growth (UN median population projection) even if all loss of farmland can be stopped.

That is a very difficult "if" indeed, because population growth consumes the farmland it needs. Nonfarm uses—urban sprawl and the needs of growing populations for space, the demands of transportation, industrialization—eat away at arable land. The losses are sometimes masked by the conversion of forests and range to farmland. Brazil, for instance, has almost doubled its arable land since 1970, but at a high cost in forest destruction. Indonesia's cultivated acreage is up 23% for much the same reason. One may be pretty certain that acreage gained in that way will be poorer farmland than the acres lost around the cities.

Erosion, salinization of fields and desertification are more serious problems in most countries than loss to nonfarm uses, and all three processes are worsened by efforts to use the land more intensively than it can bear.

Globally, erosion is the most devastating of these three processes. It reduces crop yields, impairs the ability of the soil to absorb water, washes away nutrients and forces the use of more chemical fertilizer, and leads to various offsite losses, including eutrophication of wetlands, siltation of water channels and lowland flooding. The most recent Department of Agriculture (USDA) survey puts erosion in the United States at about 13 tons per hectare per year. Perhaps one ton of that soil is replaced by natural processes. Estimates of the losses in the less developed world are more than twice as high, and the horror stories from regions like Africa suggest that the estimates are conservative. Numbers like these are incomprehensible. To make the point clearer: Iowa has lost about half its topsoil in 150 years of farming, the Palouse region of the Northwest almost as much.[7] Those deep soils can still produce, but such losses on thinner soils usually mean their abandonment.

Any global numbers about land are very imprecise. There simply have not been enough good surveys in most countries. There is, however, overwhelming anecdotal evidence of worsening soil conditions worldwide, declining rangeland and spreading desertification. One UN Food and Agriculture Organization (FAO) official remarked that "the soil is not now the same soil as it was in 1970."[8] Some scattered new lands are available (on the Brazilian savannah, for example), but most potentially arable land is in forests or devoted to other uses.

One can also learn something from the converse. Where the pressure of humans is lifted, for whatever reasons, the land revives, even though badly abused land may take a long time and may not return to its original state. The demilitarized zone (DMZ) in Korea, created at the close of the Korean War, has gone back from naked hillsides to hardwood forests. Trees and wild animal populations returned to the fenced zones between the two Germanies during the cold war. Even around Chernobyl, where 100,000 people were evacuated after the nuclear plant meltdown, wild animals have multiplied despite the radiation levels.[9]

Competition with Forests

Farmland should not be made from natural rangeland, as the United States discovered with the dust bowl. It can be converted from forests, but that too has severe consequences. Norman Myers, a long-time student of forest cover, says that "Forests once covered more than 40% of Earth's land surface, but their expanse has been reduced by one-third. The most rapid decline has occurred since 1950—tropical forests have lost half their original expanse in the past 50 years, the fastest vegetation change of this magnitude in human history."[10]

The loss of tropical forests is driven by population growth. Part of the problem is imported: logging roads are opened because of the demand for timber, and settlers come in on the roads. More of it is indigenous. Slash-and-burn agriculture was traditional in the tropical forests, but traditionally the farmers would clear a small opening in the forest, farm it for two or three years and then move on when the soil productivity declined or weeds took over. There was enough forest so that the disturbed plot could grow back before the farmers returned. As their numbers increased, their need for land increased, the cycle shortened, and the forest had no time to recover. This is, dramatically, the difference between a sustainable system and an unsustainable one. Much of the land has become useless for farming and has been abandoned, but the soil loss sometimes means that even a half-century later the forest has not returned.

The loss of forests is not simply the loss of wood. Agricultural productivity downstream declines because streams flow more erratically. With firewood unavailable, weeds and animal dung are used for cooking food, impoverishing the soil to which they used to return. Beyond that are the loss of species and the release of carbon, which hastens global warming.

Ideally, most of the farmers would leave the forests, others would practice less destructive forms of agriculture, and solar cookers would save them firewood, but the demographic pressure that drove peasants into the forests makes it hard to get them out.

There have been efforts to mobilize peasant labor for reforestation. They have had varying success. The peasants are likely to cut down the trees too soon, for firewood. The efforts are worthwhile in the face of the catastrophic loss of the forests, but the driving force in that catastrophe has been population growth. The farmers and the forests were in a balance of sorts, and it was destroyed.

To rephrase my thesis: the central problem is that population growth destroys arable land and forests even as it intensifies the human demands upon them. The pressures on the two systems interact to the disadvantage of both.

Competition with Industrial Uses

Agriculture competes with industry for arable land. Already, methanol from cane and corn are potential competitors with gasoline as fuels, and "diesel oil" can be made from soybeans or vegetable oils, though they are still competitive only in special circumstances.

Genetic engineering is finding ways of substituting crops for industrial products. At one company, scientists are developing cotton that mimics cotton–polyester blends. Arable land has for centuries produced fiber as well as food; this new discovery could shift the production of "miracle fibers" from factory to farm, imposing a task on arable land that had been shifted partly away from it. In a time of pressure on arable land, that is bad news.

Other projects involve growing crops of biodegradable plastics, industrial lubricants, soaps, detergents and pharmaceuticals. Particularly if energy costs rise, these remarkable crops may be cheaper than similar products manufactured by traditional industrial processes.[11] Such developments are usually hailed as good news. They would be if they did not increase the competition for land needed to feed a growing population.

Industry competes for the forests too. The mining down of fossil fuels, particularly petroleum (chapter 3), will increase the demand for cellulose (from trees) as a substitute for hydrocarbons in the plastic industry. The industry started with celluloid and rayon; it may be heading back that way.

The Limits of Earth

How much can the Earth tolerate? In the face of very poor data, let me offer some estimates. A United Nations Environment Programme (UNEP) study, the most detailed effort so far, compared the condition of specific acreage at the close of World War II and in 1990 and concluded that 1.2 billion hectares of fertile land had been "seriously degraded" in that period.[12] That is nearly one–fourth of all arable land and pastureland.

Gretchen Daily has come up with a somewhat higher estimate: 43% of all the Earth's vegetated surface has been degraded to some degree since 1945. More than half of that degradation has led to a "great reduction in agricultural productivity," recoverable "only with considerable technical and financial investment" or not at all. She projects further degradation of something like another 20% in the coming 25 years if, as she anticipates, degradation accelerates.[13] These losses include soil loss on farmlands, the decline of crops, livestock forage and fuel biomass on arid lands, and tropical forest destruction. Any such estimate is of course only a very rough approximation. Consider it as such, and take note of the direction.

Another measure was developed several years ago at Stanford University, where a group undertook to calculate the proportion of "terrestrial net primary productivity" (NPP) from the Earth's land area being appropriated or made less productive by humankind. (NPP is the energy generated by photosynthesis from sunlight, the basic energy source of life.) Their estimate: 41%.[14] Be it noted that, in those calculations, humans must divide our share of that energy with the agricultural pests, rodents, insects, pathogens and other creatures that live off us and our works. Nevertheless, if only as a rough indicator, the estimate is a useful reminder that we may already be using or misusing two-fifths of that basic energy. It also sets a limit of sorts: what happens when the human population passes 2.5 times the present number and is appropriating something like 100%?

We are caught between a rolling growth of human population and clear signals that, on an Earth that cannot grow, natural systems are nearing the end of their capacity to accommodate that growth.

The above measurements are about the land itself; they say nothing about the extent of the damage to forests from air pollution, the eutrophication of wetlands, or the effects of ocean pollution, ultraviolet radiation, climate change or acid precipitation. We will come to those later.

THE BASIC FOODS

The Deadly Potato

I will begin with the story of the Irish potato famine. It bears repeating as a warning for overconfident moderns.

When Jonathan Swift wrote his terrible satire *A Modest Proposal* in 1729, he described the wretched condition of the Irish peasantry, reduced to serfdom by Norman conquerors in the twelfth century. He described abortion and infanticide as commonplace out of economic desperation. (I would add "in that staunchly Roman Catholic country." The Vatican could well consider where its present views on family planning can lead.) He estimated the population at 1.5 million.

Shortly thereafter, the "Irish potato" was introduced from the Americas. It found the Irish soil and climate congenial. Farmers shifted from grains to potatoes. They were the "green revolution" of their time. The increased food production led not to better conditions but to population growth. Between Swift's time and 1845, it quintupled to about 8 million. Not a remarkable rate of increase by current third world standards; it works out to 1.45% a year. (By comparison, the current rate of growth for sub-Saharan Africa is over 3%.)

The fungus *phytophthora infestans* eventually made its way from America to Ireland, where it too thrived in the cool, wet weather. In the wet years of 1845 and 1846, it destroyed the potato crop. By 1850, some 2 million people had starved and 2 million had fled Ireland, mostly to America. The population dropped to 4 million. It is just 3.5 million even now (excluding Northern Ireland) because of continuing emigration and a recent decline of fertility to about replacement level. The Irish have learned their demographic lesson, but slowly.

Centuries of mismanagement left Ireland's arable land in terrible condition. A memorable scene from Robert Flaherty's 1934 documentary *Man of Aran* showed a farm family laboriously digging the soil out from between coastal boulders with trowels and carrying it in wicker baskets to put on their field. That has changed; Ireland has now achieved very high food production, but on acreage that has declined by one-third even since 1970. Its tired soils require some of the heaviest fertilizer use in the world: seven times the rate in the United States, more than three times the average for Western Europe.

The Irish potato famine was a searing experience for scientists. The official seal of the American Phytopathological Society displays the spore of that fungus, but the society drew the wrong lesson: that science must learn to control such pathogens. A more fundamental conclusion would be that we must control our food needs so as to avoid having to become so dependent on one crop.[15] Reliance on monoculture (a single cultivar planted continuously) is the way into a trap.

So much for one early experiment with monoculture. Much of our modern success in raising yields is linked to the promotion of monocultures such as "corn-on-corn" farming in the American Midwest. Of course we can control it now. We think . . .

We will revisit the potato and its fungus.

After that sobering reminder of the fallibility of human solutions, let us ask whether future growth in food production will keep up with the demands of rising populations.

The Nitrogen Fix

A Canadian professor named Vaclav Smil asks a fascinating question: What was the most important invention in the twentieth century? His unlikely answer: the Haber–Bosch process of synthesizing ammonia and therefore nitrogen, introduced in 1913.

Nitrogen, Smil pointed out, is essential to chlorophyll, DNA (deoxyribonucleic acid), RNA (ribonucleic acid), proteins and enzymes—in short, to life itself. There are only three natural atmospheric processes for converting atmospheric nitrogen, which is locked up in N_2 molecules, into a form usable by plants and animals: falling meteors, ozonization and ionization by lightning. The only biotic process is the creation of ammonia by various bacteria, most notably by the *rhizobium* bacteria that live symbiotically on the roots of legumes (such as peas and beans).[16]

Nitrogen and water are the two most important limiting factors in plant growth. Until the Haber–Bosch process, humankind could obtain nitrogen only by green manuring with legumes and recycling nitrogen by spreading animal manure, guano from seabird rookeries and nightsoil.

Yields per hectare are the critical issue. By and large, the land that can be farmed is being farmed. We have seen that arable acreage is headed downward. Food yields must more than match population growth just to maintain a constant level of nutrition. That is where synthetic ammonia comes in. Its invention permitted agriculture to support an unprecedented population. By Smil's calculations, synthetic fertilizers now provide the nitrogen for about half the annual global crop harvest. (That is the significance of the soaring "fertilizer" line in figure 1–3.) Smil states the point more starkly: Were it not for synthetic fertilizer, the richest nations could probably get by with a change of diet, substituting cereals for meat, but perhaps one-third of the people in the more crowded and land-poor third world countries would starve. Moreover, "virtually all protein needed for the growth of more than 400 million babies to be added in the eight populous but land scarce countries during the 1990s will have to come from synthetic nitrogen." (The eight countries are China, India, Indonesia, Bangladesh, Pakistan, the Philippines, Egypt and Thailand.)

The Age of Imbalance

This dependence is absolute. With nitrogen we stand, like Thoreau's Indian, on the inelastic plank of famine. If more nitrogen cannot sustain the growth of yields, our only choice is to reduce the demand for food—voluntarily or otherwise.

The dependence is particularly dangerous because of something called the logistic curve. The idea is simple. Change tends to happen at a shifting rate. An innovation is introduced slowly at first. Then it takes off for a while. Finally, it slows down again as its potential is exploited. The history of airplanes is a good example: the 50-year period from the

Wright brothers to the development of commercial jets was one of accelerating and breathtaking change; the subsequent 40 years has been marked only by minor tinkering, and our attention has shifted to the new technology of rockets and space. The internal combustion automobile engine is another case in point.

The history of corn yields in the United States is an excellent example of a logistic curve. It is also an important one because U.S. corn production supplies feedgrains to the country and much of the world.

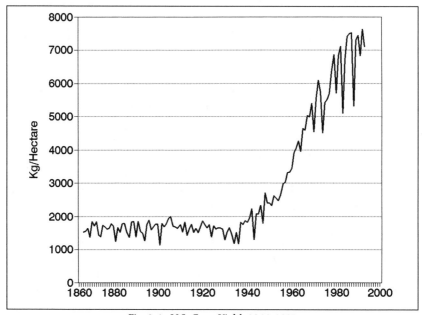

Fig. 2-1. U.S. Corn Yield, 1866–1993
(*USDA / Worldwatch, 1995*)

I can think of no technology that has escaped the logistic curve in some form. The only escape is to keep finding new technologies, and one of the great unanswered questions of our time is "can we keep it up?" An FAO official listed "the plateauing of yields of high-yield varieties" as the principal food supply problem right now, along with land degradation.[17]

We are in the era of the declining response curve. Adding progressively more fertilizer produces diminishing increases in food output. Eventually, it produces a zero increase.

Discoveries are changing our lives. In the key area of food production, however, they are refinements, not breakthroughs. We are near that last stage of the response curve unless some new technology emerges, and there is none in sight other than genetic engineering. At best, its contribution is a long way off; at worst, it may be very limited (p. 32). Population growth is with us now, and to argue the hope of salvation by technology is an imprudent policy indeed.

For a time, chemical fertilizers led to lower prices by sharply increasing food output. Eventually, diminishing responses and the rising curve of demand will raise food prices. Those who can afford to will pay those prices even if they must divert income from other uses to food. Higher prices will justify the use of more fertilizer for a time, but it will not produce much. The absolute limit will be reached, of course, well before the response reaches zero. When the food output doesn't pay for the fertilizer, food production hits a wall.

The developed world is very close to that wall. U.S. fertilizer use per hectare is only one-fourth of that in Japan, because Japan maintains rice prices five times the world level through import controls and subsidies. Our average cereal yields are 85% of those in Japan, even though we grow cereals in generally drier regions. The Japanese didn't get much for that extra fertilizer. Both we and they know that; fertilizer use in both countries has been declining for years.

The world's use of fertilizer seems to have peaked, and the world curve of grain production has turned down (figure 1–3). Whether temporarily or permanently we do not know. Further growth will depend on whether other, usually poorer nations can afford to play catch up. Indeed, they could profitably use more commercial fertilizer, within limits. They would, however, be well advised to take Vaclav Smil's calculations seriously. If they embrace commercial fertilizers as a substitute for controlling population growth, they will wind up in China's population trap (see chapter 7).

The Costs of High Yields

The reliance on chemical fertilizers has other costs and consequences even more significant over the long term than the diminishing response curve.

The Soil. One by-product of commercial agriculture has been the loss of organic materials in the soil that maintain its productivity and help to control erosion. In traditional agriculture, organic tilth was sustained by mulching, green manuring and manure spreading, as organic material was removed in the form of crops. Properly managed, higher production can actually be a help by producing more biomass (the weight of organic material grown). But hybrid corn and some of the green revolution crops make the loss of organic material worse, by increasing the proportion of harvested cereal and reducing the share of stalk, roots and leaves that stay in the soil. "Organic farming" is now enjoying something of a rebirth. It is an effort to go back to some of the old ideas, but it alone cannot maintain the short-term yields achievable with synthetic fertilizers—which means it cannot support as many people except in local situations where sufficient biomass can be applied from nearby sources.

That is the synthetic fertilizer trap. With 6 billion people on Earth, we cannot go back to techniques that supported 2 billion.

Monocultures. The pursuit of high yields leads to overwhelming reliance on a few cereal strains and to continuous planting of them rather than crop rotation. This practice depends upon synthetic fertilizers. Limiting the number of plant varieties and continuous cropping are both invitations to the proliferation of crop pests. This leads to the intensive use of pesticides. The fertilizer and pesticides in turn cause water pollution and threaten fisheries and other species (though the problem is mitigated if the pesticides are highly selective and quick to break down).

Pests, Weeds and Mutations. A more ominous threat is that pests mutate and become resistant to known pesticides. The very process of fighting insects and other pests has a frightening side effect: it promotes the evolution of resistant strains and species. The mechanism is fairly straightforward: create environmental conditions that are ideal in all respects except one—for example, a field full of corn, replanted annually, as seen by a corn borer—and there is a tremendous reward to the organism that learns to handle that one negative factor, the pesticide. That is what leads to successful mutations. Successful, that is, from the standpoint of the pest.

The more intensive the agriculture, the more likely the consequence. If farmers have no effective defense, a blight can lead to the loss of an important fraction of food supplies. The United States narrowly averted such a corn blight in 1980. The fungus that led to the Irish potato famine is endemic in the United States. Aggressive and resistant forms appeared in Europe in the 1980s and the United States in 1990.[18] Either the corn or potato blight, or something else, could surface in a more virulent form.

Losses to pests have risen despite massive increases in pesticide use in the past two generations. David Pimentel and his associates estimate that in the United States, total crop losses from insects doubled between 1945 and 1989 to 13% despite a ten-fold increase in the use of insecticides. Losses from all pests, including plant pathogens and weeds, are estimated at 37% of total potential food and fiber crops. The pesticides, aside from promoting resistant pests, destroy the natural predators that could control pest outbreaks, and they kill bees that are important to crop pollination.

Add to those losses the economic costs of pesticide-related human illness, estimated at about $787 million each year in the United States.[19]

I have been citing the United States simply because more studies are available than for countries where the problems may be even worse. In some third world countries, farmers still lavishly apply DDT and even dump it in streams to kill fish, which they gather and sell for food—DDT and all.

Weeds are on the rise. Like insects and other plant pathogens, they seem to be recovering from our chemical onslaught, and this sets another limit for rising yields.

Despite the penalties, the dependence on pesticides and herbicides has risen dramatically. Farmers are on a treadmill because they would lose part or all of their crops if they stopped spraying, given the decline of natural predators.

This is a picture of agriculture out of balance. It can be put in balance only if productivity is enough higher than demand so that farmers will move back to more benign but less intensive agricultural practices. And that shift gets harder and harder as population grows.

Energy. Synthetic fertilizers use fossil fuels for energy and as a feedstock. The additional energy needed to produce more fertilizer intensifies atmospheric pollution and the greenhouse effect. These problems will probably get worse. Right now, fertilizer plants use natural gas as both feedstock and energy source. Gas is presently the coming energy source, but its time is limited as it supplants petroleum (p. 50). The next choice for a feedstock is likely to be coal, leading to more pollution. In the process, higher energy costs (chapter 3) will make fertilizer more expensive—moving that "wall" (p. 25) closer to us.

N_2O and the Atmosphere. Some 20 to 40 million tons of nitrogen escape annually from the world's croplands into the atmosphere as nitrous oxide (N_2O). Commercial fertilizers account for perhaps 40% of that release, depleting stratospheric ozone and contributing to the greenhouse effect.

This is another manifestation of a problem that will reappear through this book: one disturbance—in this case, human tampering with the Earth's nitrogen budget—generates a ripple of disturbance.

■　■　■

What does this all mean? The world pays a high price for abandoning diversity for high yields, which in turn is forced by rising demand, driven by a growing population.

Solutions beget problems.

The 50-year rise of yields is slowing or ending, and the world is paying a high and rising price for the effort to keep raising yields. Countries that have become dependent on high yields should be seeking to escape the squirrel cage of rising demand. Countries that are not yet hooked on commercial fertilizers should recognize their potential limits and costs and look to controlling demand—population growth—rather than hoping to rely on higher food yields to solve their problems.

How much help could conscious population policies realistically offer? In theory, if the world were one market, and if somehow the "low" population projection could be achieved, there would be just 69% as many people to feed in 2025 and 37% as many in 2050 as compared with the "constant fertility" projection. This doesn't mean that the low projection would take the pressure off world agriculture. The low projection for 2050 would still be one-third higher than present world population.

In fact, most of the difference between the two projections is concentrated in the poorer countries. They are unlikely to be able to feed the numbers suggested by the low projection, and they cannot hope to feed the populations projected in the constant fertility projection. I will disaggregate the analysis in parts II and III and suggest some very different connections between food and population in poor, emerging and old industrial countries.

WATER

The world's supply of water is nearly constant, but we are making it much less accessible and usable for our purposes and other species'.

Fertilizers and Water Quality

High-yield agriculture pollutes the water it uses. Almost as much of the nitrogen on croplands is lost to erosion and leaching as is taken up by crops. Along with phosphate fertilizers, this causes groundwater contamination and eutrophication of streams, lakes and wetlands in the industrial countries. (For some particularly frightening European data, see chapter 7.)

Since wetlands are the nurseries of many ocean fish, this process—along with overfishing—hastens the decline of world fisheries. By flushing nitrates, phosphates and pesticides into the sea, agriculture may also be responsible in some way for a series of observed ocean changes such as a 70% decline in the sampled volume of zooplankton in the California current since the 1950s,[20] the worldwide decline of pelagic mammals and coral reefs, the spread of poisonous "red tides" along U.S. coasts, and the loss of porpoises along U.S. coasts and harbor seals off the coasts of northern Europe. Those mysterious changes may also be related to climate change or to some other human activity; we do not know.

Water as a Limit

Arable acreage depends on rainfall or irrigation. Acreage has reached its limits in most of the world where there is enough rainfall, but irrigation has added acreage in drier climates and has played a central role in raising yields. High–yield cultivars usually require a lot of water in order to utilize heavy fertilization. This frequently requires irrigation. Water use has risen faster than population growth, but the world is pressing against fixed limits. In another of the series of studies of humankind's pressure on "renewable" resources, Sandra Postel, Gretchen Daily and Paul Ehrlich have undertaken an estimate of how much of the accessible runoff is appropriated for human use. Their estimate: 54%. They note that new dams could add perhaps 10% to the total accessible water, far less than population growth in the next generation.[21] I would add that those dams would displace people and flood farmland. You can't just do one thing.

After drawing first from surface stream flows, irrigation becomes increasingly dependent on groundwater. It does not take many decades to overcommit the groundwater resources. Data on water availability and groundwater replenishment are notoriously uncertain. For the United States, some experts believe there are no reliable estimates. Others have offered estimates from 10% to 25% for the proportion of annual groundwater consumption that is "mining" the resource—using it faster than it is replenished. Water tables are falling in some of the major aquifers, but there is no systematic overall data collection as to where or how fast. As water is overdrawn, it often becomes so mineralized as to be unusable, or the water table gets so deep that it is uneconomic to exploit.[22]

Water availability varies tremendously. Canada has perhaps 106,000 cubic meters per capita available annually and uses only 2% of it. The Persian Gulf states and Libya have almost none and rely on desalinization for drinking water. Most of North Africa and the Middle East have less than they need, and population growth will make the problem desperate in the next generation (chapter 8.)

The Soviet Union's "New Lands" project used water from the Pamirs to produce cotton, and they pushed the system beyond the limits. That

0944499

dried up the Aral Sea, which became a desert disaster of blowing dust laced with residual pesticides, with abandoned fishing boats resting 100 kilometers from the present water's edge. The International Rice Research Institute warns of coming water scarcities in Asia and estimates that Pakistan, saddled with a century-old irrigation system that cannot be flushed, has suffered salinization on 10% of its irrigated land.[23]

We shall see in chapter 4 that global warming, coupled perhaps with rising air pollution, is expected to make the arid parts of Earth even drier.

This is another of the limits against which humankind is pressing. The kindest generalization is that there will be less water in the future where it is most needed—and much less per capita.

Humans tend to press against the margins. Rain-fed agriculture expands in moist periods without regard for climate variability. There have been recent droughts in many parts of the world—England, Spain, northern Africa, southern Africa and north China (even while there were floods in south China)—and prospects for more of the same with climate warming (chapter 4). Most of the richer nations can get by so long as there is food available in world trade—though I will show in chapter 8 that this is not much of an assurance. The poorer countries are not so fortunate.

Of course water can be used more efficiently. Much irrigation water is "wasted" in much of the world. Put it in concrete ditches or pipes and you will have more for agriculture—but you will dry up gallery forests and stream valleys and reduce the natural complexity of nature. Is this a desirable human goal? Or would we be better off limiting the demand side and not appropriating all the water?

Irrigation is expensive and often subsidized. An acre of irrigated corn in Nebraska requires about three times as much energy as an acre of rain-fed corn.

There is no shortage of salt water, but desalinization is an order of magnitude more expensive than irrigation.[24] It can cost 14 times as much as irrigation, including the energy cost of moving the water from coasts to fields. There may come a day when desalinization becomes a realistic possibility, but although the search for cheaper desalting processes has been intense for two generations, nobody has shown the way to a breakthrough. That day when crops are raised on desalted water will be a time of desperation, when food prices have gone so high as to justify the energy cost and additional pollution. As energy prices rise over time (chapters 3 and 4), they will drive the price of this alternative even higher.

Because of the limits on finding new water, some investigators are turning away from the search for water and instead looking for drought-tolerant crops, such as the desert jojoba for oilseed or drought-tolerant varieties of wheat. In effect, the proposal is to expand acreage by finding crops that can grow in semidesert or desert conditions, since the prospect for higher yields is not encouraging. So far, this line of inquiry has led to more hopes than results.

In another approach, agronomists have sought to develop salt-tolerant crops and even to identify useful plants that can be irrigated with seawater. Again, these "miracle crops" have been disappointing.

To put it another way: here is another demonstration of the rising cost curve when we try to push the exploitation of Earth's natural endowment too far.

Agriculture is the biggest water consumer in irrigated areas, but it has a low priority. People will outbid it to purchase water for drinking water and industry or even golf courses, swimming pools and lawns. "Water ranching" in the southwest United States consists of buying up water rights from farmers to support growing human populations. In California, the laws have been changed to make it easier for cities to acquire water supplies once reserved for agriculture.[25] The priorities may change as the situation tightens. There is something basically out of shape about a process that brings in more people and more demand as it wipes out the food production to supply them.

■　■　■

In short, water is part of the problem, not a solution. Added to the other limits described earlier, the limits on water availability will constrain hopes for rising food production as they never have before.

TECHNOLOGY AND FUTURE FOOD NEEDS

Our society is addicted to technology. What lies out there that might help supply this most basic human need? And how can we escape the Irish experience with the potato?

The Haber-Bosch process did for agriculture in this century what the harnessing of mechanical power did for the Industrial Revolution in the nineteenth century. Both depended on fossil energy, and both set immense changes in motion, not all of them beneficent. The Haber-Bosch process is nearly a century old. What if any technology is in the making that will have consequences equally profound?

Perhaps the two seminal discoveries of the past half-century are the semiconductor and DNA. Their potential impacts are still unfolding. The computer revolution, vast though it is, has not demonstrated more than a marginal relevance to increasing agricultural yields.

Genetic engineering may prove more relevant. It is one of the hottest fields of science right now. There are high hopes for its use in agriculture, but so far it has been used mostly to improve the appearance of foods, to make them easier to ship or to develop plants that will incorporate their own pest protection. *Bacillus thuringiensis* is a natural pesticide. It has been implanted into cotton, corn and potatoes. It is potentially valuable but hardly a permanent solution. Critics have pointed out that pests are already developing resistance to it.[26] As pests adapt, such technologies simply help us to hold our own in the war against them rather than offering the prospect of gains in yields such as those we have witnessed in the past half-century.

The Holy Grail of modern plant genetics is the use of gene splicing to develop nitrogen-fixing cereals. Legumes are the traditional source of nitrogen, but they have limits: low yields and low digestibility as a staple diet. Self-fertilizing corn or wheat would avoid those limits and bypass the need for synthetic nitrogen fertilizers. If such genetic engineering worked, it would indeed be an epochal achievement.

I offer two warnings: (1) it is a long way off, and (2) it may never help. There are some interesting experiments under way, but nobody pretends that they are close to practical application. The time between laboratory and widespread field use is particularly long in agriculture because of the need to test the experiments, to develop cultivars that work in different climates and to produce stocks of seed. Perhaps in 40 to 50 years?

Moreover, fixing the nitrogen is only part of the solution. If natural processes replaced chemical factories, there would indeed be cost savings and less pollution, but two very large questions remain. First, what yields would the new plants produce? If they do not increase the yields that have been achieved with synthetic fertilizers, they will have not addressed the race between population growth and food production.

Second, what would they do to the biosphere? I have already discussed the environmental dangers that result from human activity in making ammonia and thus releasing accessible nitrogen into the environment on a scale comparable to natural processes. Developing plants to produce nitrogen could be as dangerous or perhaps more dangerous because it might lead to even larger releases of nitrogen. If the nitrogen-fixing gene should spread into "weed" relatives of cultivated grains, there could be a green revolution among weeds, with consequences still incalculable.

Taking these reservations and inevitable delays into account, the common wisdom among agricultural geneticists is that there is no technology in sight that will maintain recent progress in yields. A Nobel laureate says,

"The human race now appears to be getting close to the limits of global food productive capacity based on present technologies . . . [Many agricultural experts] are desperately worried about the food problem."[27]

Hydroponics is sometimes offered as a solution to the food problem. Specialty items such as tomatoes have been raised in greenhouses, with high yields. Clearly, they are not competitive at this time with more traditional agriculture except for such specialized uses as offering fresh vegetables in winter. They are hardly a substitute for cereals, but the example has stirred some imaginations.

Unlike genetic engineering, the technology for hydroponics is largely in place. The principal problem is cost. It has been said that mankind could grow oranges on Mount Everest if we wanted to enough—but they might cost several million dollars each. Partly closed systems, recycling much of the water and some of the nutrients, would help to push back the limits I have been describing, but there is not enough experience to evaluate the costs of the technology. Such systems might be useful for, say, a group of scientists or technicians in a remote desert location. As a way of feeding growing billions of people, it justifies the same answer as do proposals for sending population growth off to Mars: "Have you considered the per capita costs?" The proposal strikes me as pretty unrealistic compared with the alternative: find ways to encourage people to have fewer children.

■　■　■

I raise these questions not just to be gloomy but to make the point that there are real, practical limits on grandiose dreams of limitless growth in food production to feed a perpetually expanding population.

Some very optimistic projections appear from time to time, suggesting that world agriculture can support some multiple of present population. I suggest you check their assumptions before you take them too seriously. They generally are based on the assumption that all potentially arable land is used for agriculture. Take the problem of forests: an FAO study in 1980, for instance, simply assumed that forests would be put to the plow wherever the land could be plowed, without considering the consequences, which would be immense and detrimental to agriculture itself.[28]

Those projections also tend to assume the achievement everywhere of yields reachable under ideal conditions. Usually, they assume high inputs of synthetic fertilizer without regard to the nitrogen trap that Professor Smil has so eloquently described. They do not quantify the increased demands for water or show where it will come from. And they assume that agricultural pests and weeds can be brought under control and kept that way. A bit more caution is in order.

I wrote earlier of a "wall." Perhaps a better analogy is a rapidly steepening slope. There will be more discoveries, and they will offer the prospect of better yields in one or another crop and some hope to poor nations. Much of third world agriculture does not yet use known technologies. There is room to apply them and thus increase yields. For nations at the beginning of the response curve, an investment in synthetic nitrogen is justified, if it is coordinated with steps to stop the growth of demand rather than trying to keep up with it. However, even with the low population projection, it would take an increase of one-third in yields in the next 30 years, worldwide and evenly distributed, to avoid a decline in per capita consumption.

We have an imbalance, which is the sudden multiplication of human numbers. We have tried to take care of the imbalance by creating another one: the artificial introduction of huge amounts of accessible nitrogen into the biosphere. This in turn raises the question, What problems will that solution generate? The way out of the problems, of course, is to try to get back into balance with a finite Earth, not to overwhelm it with dazzling technologies whose consequences we do not understand.

THOMAS MALTHUS, AGAIN

The modern population debate really began in the 1790s with essays by the French writer Condorcet, who thought human ingenuity was capable of solving all problems, and the English clergyman Thomas Malthus, who said that "the power of population is indefinitely greater than the power in the earth to produce subsistence for man. Population, when unchecked, increases in a geometrical ratio. Subsistence increases only in an arithmetical ratio."[29] (In his later writing, he amplified the phrase "when unchecked" by pointing to the role of deliberate birth control in avoiding that fate, but he is remembered for that stark assertion.)

Condorcet is pretty much forgotten. The poor man died miserably in the French Revolution. Malthus is still being debated, with Condorcet's intellectual heirs saying "look; it didn't happen" and neo-Malthusians saying "maybe not yet."

The difference of opinion is rooted in a difference of time horizons and in the role of technology. Neo-Malthusians generally would admit that Malthus envisaged a closed system and did not see the tremendous importance of the New World in offering new lands, accepting migration and taking the pressure off Europe. His second failure was his linear arithmetical model of food production. He envisioned growth in output only by opening new land. In an era when yields varied mostly with the weather and the quality of the land, he did not foresee that technology, cheap energy and fertilizer could raise yields geometrically. If science had not succeeded in raising yields as it has in the past half-century, nobody would be questioning Malthus now.

Despite those faults, he identified the issue. The central questions for humankind for the next few decades are likely to be which of the two geometric growths—yields or demand—will continue and how fast. Malthus recognized the policy option: limit human population growth if the growth in agricultural yields threatens to fall behind. And that is precisely where we are now.

THE LIMITS OF CONSERVATION

Whenever there is a problem, there is somebody to offer conservation as the solution. Whatever the context, those who would rather not face the prickly issues of managing human fertility argue that we can solve the problem by conservation. They argue that "food is not in short supply; it is badly distributed."

They do not answer several questions. First, how do you propose to bring about this egalitarian distribution? Second, is redistribution a better objective than higher per capita consumption? Third, even after reaching that happy egalitarian state, would you not have to advocate an end to population growth to keep intake from decreasing?

World production of cereals works out right now to about 333 kilograms (kg) per year per person. At the low edge, where all grain is consumed directly, people require nearly 200 kg, and this allows little or no meat, milk or eggs. It is exactly the sort of diet from which the third world is trying to escape. Indians consume about 200 kg, Chinese about 300 kg. Present world population, if fed at Chinese standards, would consume substantially all the available cereal if somehow all differentials in consumption were erased. Given the prospects for yields, there is little room for maintaining even the Chinese level of consumption with a growing population.

Even the richest cannot eat much more food than the poorest. They eat more meat and thus divert grains from human to animal food, but there are limits.

We in the United States would be healthier if we ate less meat and more cereals and vegetables. (How much farther we will willingly go in that direction is another question.) We consume about 800 kg of food and animal feed per person each year. Let me run a rough calculation. Our population is heading past 400 million around 2050. We currently export about 100 million tons of cereals, but we will be eating those exports by 2050 if our consumption habits stay as they are, assuming that U.S. cereal production has indeed flattened out. If we gradually cut our per capita meat consumption in half (as measured by feedgrain use) and eat cereals instead, we could export declining amounts—about 70 million tons in 2050 and much less thereafter. That is about 7% of present third world cereal production. The problem is that third world population will have increased by anything from 50% to 133% by 2050 (using UN low to high projections—and assuming no population crash).

These are horseback calculations with many uncertainties, but they make the point that even the United States, the major world supplier, can play only a marginal and temporary role in the third world food balance. These countries must solve their problems at home. In the face of plateauing yields, that means a population policy. As for the United States, I will later argue that a race to maximize yields is not a valid objective. We should be heading toward a sustainable organic agriculture, probably with lower yields.

Where does that lead us? Don't rely on conservation alone.

MEAT

One gets into some very shaky data in dealing with livestock and meat. The statistics say that permanent rangeland and pasture constitute 25% of the Earth's land surface and arable land only 11%; but meat production comes from more sources than that acreage. The drier lands are grazed—in the western United States, across Africa on both sides of the Sahara, Asia Minor and Asia as far as Mongolia and in Australia—where rain is insufficient to support intensive agriculture or forestry. In traditional systems, herdsmen may have the right to graze the land between crops or under olive groves.

Ideally, such grazing would be a beneficial use of otherwise unproductive land, but reality is far from the ideal. This is a fragile system, easily weakened and slow to recover. Even before human populations took off on their current spurt, there was a tendency for pastoral peoples to push beyond the limits of their support system, which led to chronic overgrazing everywhere. The goat has long been recognized as the scourge of the Middle East; the dramatic recovery of managed systems where goats are excluded testifies to the damage they do.

Modern grazing practices are not much better, at least in the United States. Some 68% of Bureau of Land Management (BLM) rangeland is in less than satisfactory condition even by the management's own assessment. General Accounting Office (GAO) reports of overgrazing and range deterioration are even more critical.[30]

Population growth has been worsening the damage even as it increases the need for grazing. The drought of the early '70s in the Sahel depleted the livestock. Within a few years, the number of animals had come back up, but the number per person was far below predrought levels because of population growth even during that tough period.

Pastoral systems are already in danger, and climate change seems likely to hit the traditional rangelands hardest (chapter 4), leading to spreading desertification and the loss of livelihood for many pastoral people.

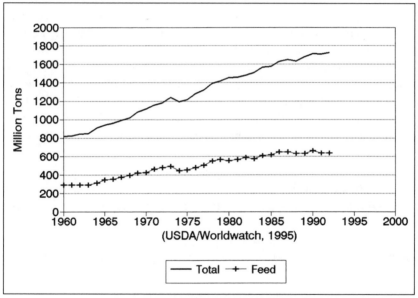

Fig. 2-2. World Grain Use, 1960–1992
Total and for Livestock Feed

The decline of grazing has made meat production increasingly a function of hay and cereal production as the animals rely more and more on feed, and this intensifies the competition for grain just when a rising human population most needs grains for food.

Rising output supported increased grain for both food and feed until the late '70s, but grain production for feed has stagnated since then.

There has been a sharp shift from cattle and sheep, both because of the declining role of range-fed animals and because pigs, chickens and grain-fed fish farms are more efficient users of grains and table scraps.

Estimated consumption of food per capita is a fairly good proxy for diet diversity, and it tracks with the indicators above. Worldwide, it rose by 8% from 1970 to 1985 but seems to have declined slightly since then (1991 estimate).

Modern chicken farms can convert two pounds of grain to almost a pound of carcass. Egg production and hog farms do perhaps half as well. The animals are treated without mercy, kept in tiny cages and used simply as digestive processors. It is hardly a pleasant thought, but this sort of agricultural factory will probably spread. Since it is feed-efficient, it will supplant more humane husbandry as cereal production loses the race against population growth. This will postpone the impact of stagnating cereal production, but not forever. It is hard to imagine that technology will be able to carry the efficiency of conversion much beyond 1:2 (one kg of carcass for two kg of feed).

Meat production is a particularly sensitive derivative of cereal production. When people are hungry, they eat the animals and then revert to a grain diet. Not a popular prospect. In many countries, the poor have enjoyed an improving diet in recent decades, but the trend may be ending. The end of a period of rising hopes is a particularly dangerous time.

■　■　■

This is an area in which limiting population growth and therefore demand would be particularly useful, but it won't happen soon. Even the UN low projection means more and more people for the next half-century. It does not drop below current population for a century. The price competition for feedgrains is tightening the entire cereal supply, at the expense of the poor.

FISHERIES

Ocean fisheries have traditionally represented a source of protein independent of agriculture. They face, not an imperiled future, but an impoverished present. We are face to face with a dramatic current example of a growth/collapse cycle. One can see the growth phase beginning to turn around in figure 1-3. Aquaculture is a form of farming rather than a traditional fishery. If we extract it from the line in figure 1-3, we get a curve looking like this:

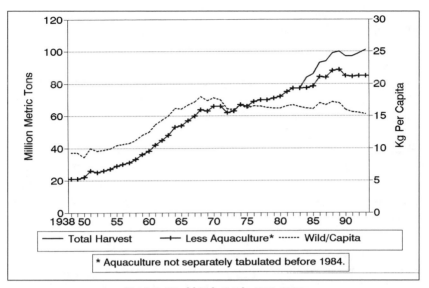

Fig. 2-3. World Fish Catch, 1938–1993
(*UN Statistical Division, FAO and Worldwatch*)

The "+" line is the annual catch of wild fish (plus aquaculture before 1984; I cannot find estimates of worldwide aquaculture output before then). The dashed line shows per capita consumption worldwide. Its recent decline demonstrates the importance of arresting population growth to preserve an important source of protein in human diets.

More and more of the catch consists of species that were once considered "trash fish." One can sing a dirge for one after another high-quality fish stock that was once plentiful and now is threatened or close to extinction: Atlantic codfish, haddock, flounder, sole, bluefin tuna, Chesapeake Bay oysters and the salmon of the U.S. west coast. There is considerable scientific uncertainty about world stocks but apparent agreement that 45% of U.S. fish stocks are being depleted and 56% of European stocks.[31] Fish catches have declined from their peak in every region except the Indian Ocean by anything from 2% to 53%. Fisheries produce some 16% of animal proteins consumed worldwide, plus oils and fertilizer. A major food source is on the decline when it is most needed.

"Independent" is perhaps the wrong word to use for fisheries. They have been particularly vulnerable to a wide range of human activities aside from overfishing. Destruction of coastal wetlands wipes out the nurseries of many species, as does eutrophication from growing coastal cities and too many humans, too much fertilizer and industrial wastes. On the other hand, fisheries flourish with a certain level of terrestrial nutrients. The construction of the Aswan Dam stopped the annual floods of the Nile, cut off that flow of nutrients and wiped out the fishery at the mouth of the Nile.

Dams and the siltation of mountain streams from logging and construction wipe out the spawning grounds of species like salmon. Climate change may affect ocean temperatures and circulation and affect fish stocks (chapter 4). Increased ultraviolet radiation from human activities (chapter 5) may threaten the plankton that supports the entire ocean biosystem. We do not know what other effects human activities are having on the oceans, but something is happening to them (p. 29). The threats to fisheries are perhaps the most dramatic distillation of the anthropogenic disturbance to world systems. Ecosystems can change and adapt, given time, but not to rapid change such as humans are inflicting on the oceans.

A declining resource leads to violence. There have been fishery "wars," involving gunfire or the threat of it, between Canada and Spain, Iceland and the United Kingdom, Norway and Iceland, the UK and Spain, and some tense negotiations among the United States, Canada and Alaska (which is behaving very much as if it were sovereign—and hungry). The Europeans and Canada have patched up their differences momentarily, but the near–hostilities suggest how serious the problem is.

It is a matter not just of protein but of labor. Fishing communities in Spain, eastern Canada, New England and the Pacific Northwest are on the dole and thoroughly angry. Partly to mollify those constituencies, governments worldwide subsidize ocean fishery to a remarkable degree. World fisheries, by one FAO estimate, yielded $70 billion in gross revenue in 1989 but cost $92 billion to operate; another estimate puts the deficit more than twice as high; it is covered by subsidies.[32]

Could we be more stupid—subsidizing fishermen to destroy a resource that we badly need?

The subsidies have supported the development of modernized "factory ship" ocean fishing fleets that have displaced small–scale indigenous fisheries. As with other technology, productivity rises, but that does not help the displaced fishermen. Fishery policy thus intensifies the world-wide problem of unemployment (chapter 5) even as it destroys the resource. Improved efficiency and rising demand are converting an ongoing source of protein into one mighty, unsustainable harvest.

The crash of fish populations is thus partly a matter of policy, and the consequence is less food even as populations need more. Ocean fishery will be saved only if nations can agree to regulate the world's ocean commons much more effectively. That will require moratoria or drastic reductions in fishing most stocks and a suddenly reduced contribution to world protein intake. It must happen soon; there is concern that some fish stocks depleted below a certain level may not be capable of recovering even if fishing stops.[33]

Aquaculture is often offered as the solution. It has doubled worldwide since 1984 and now produces about 13 million tons—two-thirds as large as the total world fish catch in 1950. It will probably play an increasing role, but it has its costs. It pollutes and it can cause eutrophication because it unnaturally concentrates fish populations. One contested estimate put the raw sewage output from a two-acre salmon pen in the Puget Sound as equivalent to a town of 10,000 humans. The crowded fish pens make fish epidemics a constant threat—is there a parallel here to third world cities?—and the use of antibiotics in the fish feed to prevent epidemics has led to the growth of resistant bacteria in nearby waters.[34]

Fishing is akin to hunting, except that it took longer to run down the stock. Aquaculture is really animal husbandry. Is it an efficient use of feed? That depends. Valuable fish such as salmon are inefficient, but the price justifies their culture. Catfish may be competitive with chickens, or nearly so. In east Asia, tilapia fish are raised in the flooded rice paddies, providing both protein and fertilizer, but this works only in limited areas.

Viewed this way, aquaculture has a role and may expand. Since the most productive aquaculture competes with other agriculture for land and most aquaculture competes for feedgrains, it should be seen as a form of agricultural diversification rather than a true substitute for declining ocean fishery, or meat production, or even cereal production.

Other Ocean Resources

Some optimists look at the ocean, see that it covers most of the Earth and say "there is the solution to human nutrition."

Seaweed has traditionally been harvested and eaten in Japan and to a lesser extent in Korea and China. It is periodically proposed to expand the harvest to take care of rising human consumption needs. I know of no studies of the potential size of the resource. The proposal, like most others that deal with human overconsumption of resources, would lead to another encroachment of human activity on other species' life support systems and a further disruption of marine life cycles. I have no reason to think that the proponents have looked at such consequences. (The whole world needs to learn to do environmental impact statements—see chapter 18.) Our species is very good at setting unpredictable chains of events in motion.

Ocean plankton offers a much vaster potential resource and poses much more frightening dangers. It is the base of the ocean food chain and supports the major fisheries of the world. Krill, one element of the plankton community, has been harvested on an experimental basis and used for direct human consumption. The proposal to harvest plankton for a significant fraction of human food needs is in effect a proposal to shorten the food chain by bypassing (and wiping out) intermediate predators (i.e., most of the world fishery) and eating the plankton ourselves.

Some would take the proposal another step. Ocean algae support the rest of the plankton. Experiments have suggested that the reason there is not more plankton is that growth of the algae is limited by the lack of a trace mineral: iron. Spread minute quantities of iron in the sea, the argument goes, and the algae will multiply, which will lead to an explosion of the plankton. We can eat the plankton or leave it to the fish, which in turn will multiply the world fishery. Moreover, it is argued, this fertilization would help to avert a warming climate. Ocean algae already absorb an unknown fraction, perhaps nearly half of, the carbon that humankind is releasing into the atmosphere. Increase the plankton, it is argued, and you diminish the problem of anthropogenic carbon dioxide (CO_2) and climate warming. Small-scale experiments have suggested that fertilization of part of the sea with iron may be possible.[35]

This sort of proposal raises a question and a warning. What would be the other effects of that fertilization? Even if we could satisfy ourselves on that question, this gigantic environmental engineering experiment would not solve the problem of human population growth. It would only defer it.

The proposal raises profound issues. It is perhaps the ultimate example of the mistaken policy that we have pursued on land. To correct two imbalances—global warming and fishery declines—they propose to set even vaster changes in motion. Fortunately, the American Society of Limnology and Oceanography has weighed in strongly against the idea[36], and the Intergovernmental Panel on Climate Change (IPCC) climatologists (chapter 4) warn that engineering "solutions" to climate warming are risky and probably not practicable.

Where do we want to go? The human experience at engineering the environment so far is approaching the disastrous. Do we want to plunge ahead and engineer the Earth on a scale even more huge? Do we want to take the risks? Is it so important that we try to accommodate, say, a doubling or trebling of the human population? I love the seashore. I would hate to contemplate our deliberately embarking on the creation of a sea without fish, without seabirds, opaque green and slimy with algae. It is a conceivable world. It isn't the world we were born into. We have already done it on a small scale. Go visit an industrial area and look at the pounded, oily earth and the ponds full of algae.

This brief excursion into futurism forces us to look at a question that usually goes unasked when such proposals are made:

What kind of world would it be if we succeeded?

The proposals, finally, force us to ask, What is optimum? If in fact we could support larger populations—at least momentarily—we can no longer leave it to Allah or God to decide how much farther population growth should go. We need consciously to seek the optimum, i.e., the levels of production and consumption that protect the resource base, preserve the complexity of the environment and provide enough food for a decent and mixed diet for all. I will return to the concept of optimum population in chapter 18.

In a sense, such futuristic visions simply sharpen the issue. We are already making such choices in agricultural, forestry and fishery policies, without considering the consequences. We are driving ourselves toward a desperate future by treating population growth as an independent variable that we must accommodate rather than a force that we can influence.

THE COUNSEL OF PRUDENCE

For all the basic elements of diet, the cheap gains have been taken. Further additions to food supply are costly and pose the threat of fundamental environmental dislocations. They conflict with other sectors by demanding increased energy input as we approach the energy transition away from petroleum, and they offer the prospect of some dismal futures. There is another basic reason for restraint on the demand side:

We must not push the system to the edge.

Climatologists tell us that the '60s and '70s were an unusually favorable period of weather for agriculture. Experience suggests that we may reasonably expect some worse weather, and indeed we may currently be getting it. The past decade certainly has not been a great one, with its floods, hot and cold waves and droughts; the occurrence of tornadoes in the United States, for example, has risen steadily and in 1990–94 was treble the rate of the early '50s.[37] Beyond that lie the unpredictable effects of human activity on the climate (chapter 4) and the likelihood of greater heat and aridity in the most vulnerable zones (p. 66).

We must expect that pesticide–resistant pests (p. 26) will score some victories. We need crop diversity so that other sources of food will be available when a blight strikes the major cultivars. Diversity probably means lower production.

Much cropland should be farmed less intensively, with a view to sustained production, or shifted out of crops to controlled grazing or forestry (p. 19). The world needs room to accommodate such changes. This requires that population and demand be brought down as quickly as possible to where the loss could be absorbed without disaster.

To quote a long-ago ad for the Volkswagen Beetle: "Live below your means."

Chapter 3

The Energy Transition

WHEN THE FIRST OIL WELL came on line, in Pennsylvania in 1859, there were just over 1 billion people on Earth. Fossil fuels, particularly oil and gas, have supported the automobile age, the evolution of modern industrial society, the rise in food production and the population explosion.

Unlike the food problem, the issue is not that we have reached a plateau. Energy is the dominant fact of our universe. Potential human needs are almost infinitesimal compared to the energy contained in a few tons of matter. The problem, rather, is how to get at the energy in an available and benign form and to recognize the limits to its safe use. As in so many areas, solutions create problems. Nuclear fusion is no exception.

The immediate issue is how to shift from fossil fuels with their concentrated energy to renewables (wind, direct solar energy, geothermal, perhaps tidal energy or ocean heat gradients), which must harvest much more diffuse energy. The world is not doing much to prepare for that transition.

CONVENTIONAL ENERGY: THE SECOND TRANSITION

From the economists' standpoint, the problem with conventional energy is that it is running out. From the environmentalists' standpoint, the problem is that it is not running out fast enough. Let us take up the economists' problem first.

"Conventional" energy and the problems it causes are really rather new. Through the middle of the last century, most energy in the United States (and probably the whole world) was from firewood.

Commercial energy use worldwide rose by 40% between 1970 and 1993. About 87% of that energy comes from the conventional fossil fuels: petroleum, gas and coal. Population grew by 50% in the same period, so energy use per capita declined slightly, because of some success in using energy more efficiently plus an economic slowdown that left energy growth flat from 1990 to 1993. The U.S. Department of Energy's Energy Information Administration (EIA) expects a 36% increase from 1990 to 2010—slightly above the expected rate of population growth—despite a sharp rise in the projected efficiency of energy use. The growth is expected to be fastest in the newly industrializing nations.

One problem is how much energy the industrial world uses. The United States consumes more than 30 times as much energy per capita as does sub–Saharan Africa. Another problem is that the rest of the world, if it could, would use nearly as much—and some of the newly industrializing nations (chapter 7) may succeed.

Petroleum

The most widely accepted projections of the total resources of petroleum discovered or likely to be discovered are done by a working group of the World Petroleum Congress, with the U.S. Geological Survey (USGS) prominently represented. The most recent projection is shown here.[38]

How long will that supply last? Worldwide, it works out to something less than 50 years of proven reserves *at current consumption levels* and 63 to

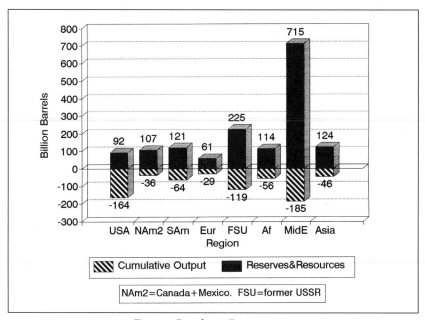

Fig. 3-1. Petroleum Futures, 1992
(14th World Petroleum Congress)

95 years, including estimated but unproven resources, with a preferred figure of 71 years.

The use of *current consumption levels* is a misleading device. If consumption is growing, the resource will disappear faster. In fact, consumption grew about 1.6% per year in 1970–1990. Projections for 1990–2010 range from 1.5% to 1.7% per year if there is a sharp rise of 40% in the efficiency of energy use in the third world, they rise to 2.5% in the absence of such improvement.[39] As I will explain in chapter 7, even the higher projection may prove conservative.

If we adjust the estimate to reflect a growth of consumption at those rates, the longevity of the resource declines from 71 years to perhaps 42 to 48 years.

These "years' supply" projections are simply a tool for visualizing the size of the resource. (*Quads* and *petajoules* are not easy for most of us to visualize.) In the real world over the long run, the rate of use will almost certainly decline as the resource peters out, and production may spin out at a diminishing rate for decades. Crude as they are, however, the calculations make the point that petroleum is a vanishing resource.

One sees glowing news reports of new discoveries. The USGS has recently raised its estimate of the total amount of oil in existing U.S. oil

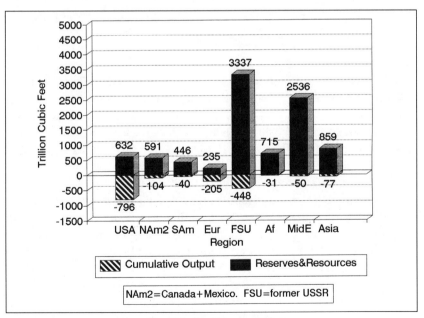

Fig. 3–2. Gas Futures, 1992
(*14th World Petroleum Congress*)

fields. Production from the North Sea oil fields has risen, and Norway is now the second largest exporter after Saudi Arabia.

Don't be misled. The rise in U.S. estimates was partly real and partly the result of changed methodology (chapter 10). Norway's golden ride will be brief; production is expected to peak and turn down about 2000. Overall, the world estimates are fairly stable. The estimates in figure 3-1 are 5% higher than those in 1991 and 5% below a 1994 UN estimate. The figure for reserves is 10% higher than that in the *American Petroleum Encyclopedia 1944*. The real moral is this: Don't celebrate; plan ahead. The trend is more important than a few years' difference in the estimates. And the inescapable trend is that we are coming to the end of the petroleum era.

The general complacency is remarkable indeed. Consider this:

> The World Energy Council . . . found that "the fears of imminent resource exhaustion that were widely held 20 years ago are now considered to be unfounded . . . proven reserves alone could supply petroleum needs for 40 years . . ."[40]

Is 40 years eternity? The Energy Council is still caught in the fallacy of "current consumption." More than that, it apparently has not heard of the concept of sustainability. It seems to me that the message should have been this: "The supply is finite. Even assuming the unproved resources are there, we have supplies for only 40-odd years. We must develop policies to adjust to that reality." Adjusting means developing unconventional energy sources, and it also means addressing demand, which starts with population.

Conservation can help to extend petroleum supplies. There are more opportunities in energy than in agriculture. Exactly how much more is hotly debated. Advocacy groups argue that more than 70% of current use can be saved simply by using known technologies such as insulation and more efficient motors. The Electric Power Research Institute sees about half that much potential savings. The IPCC (see chapter 4) sees possible savings in the 10%–30% range in the next two decades through technical efficiencies and better management and 50%–60% using the best-known technologies regardless of cost. Those estimates might be realistic for an ideal world in which everybody did the right thing. In any event, conservation is a diminishing "resource." Some gains can be made easily and are now being taken; others are problematic or require substantial changes in the energy system. Successive gains become more and more difficult, expensive and hard to find. Mathematically, each successive 10% savings is a smaller absolute amount. In short, conservation is just one transient part of any program to face the energy transition.

There are dangers in depending on distant sources. Petroleum resources are highly concentrated in the Persian Gulf area. All of the industrial countries except Norway, Great Britain and Canada are heavy petroleum importers. Many of the major newly emerging industrial countries also depend heavily on imports, and the supplies lie mostly in an unstable part of the world. It is a bit disconcerting to rely on Saudi Arabia to keep one's pipes from freezing tonight.

Instability in the Middle East—from Moslem fundamentalism or even perhaps a democracy movement—could threaten an interruption of oil supplies and lead to such desperation in oil importing countries as to force action, reluctant though it might be. The industrial nations, with the United States in the lead, would probably find it necessary to launch another Desert Storm, even if we had no Arab allies this time—and whether or not we sympathized with the politics of those who created the disruption. Such an intervention would earn the epithet "imperialist" from those at home and abroad who believe in self-determination. It would be extremely costly in economic and political terms, and it might not work. Do the industrial nations really believe we can prevent sabotage of wells and supply lines? The brief Iraq–Kuwait war suggests not. Rout though it was, we did not stop the Iraqis from flaring hundreds of wells. In another Desert Storm, we might find more of the Arab world helping to sabotage the supply.

Arab fundamentalism has been on the rise in the Middle East; unrest, assassinations and security problems have dogged the region. Even Saudi Arabia, the linchpin of the world's petroleum supply, has had to raise taxes and cut some of its domestic largesse. The Saudi clan has been buying loyalty, or at least acquiescence. The new austerity is not popular. Moreover, the clan has begun to face open criticism from some of its intellectuals for the lack of democracy. The recent terrorist murders of American advisers and State Department advisories urging caution in traveling in Saudi Arabia are jolting reminders that even Saudi Arabia is not secure.

That should remind those who are content with "40 years'" supply that we have no assurance even of that supply. The instability of petroleum supplies, almost as much as their ultimate limits, suggests the importance of moving to a more diversified energy base (chapter 4). And, since the magnitude of such a transition depends on the total demand, we need both conservation and a population policy.

Gas

The easiest diversification is to gas and coal, but neither offers a permanent solution. The World Petroleum Congress' figures for gas reserves and resources work out to a comfortable 132 years' *current consumption* but just 59 years if deflated by EIA projections of rising demand, as I did with the petroleum projections. This calculation does not make allowance for the shift from petroleum to gas as oil resources dwindle. On the other hand, estimates of gas resources, unlike those for petroleum, are still rising. Gas has the attraction of being "cleaner" than oil and easily substituted in electric power generation, but those virtues make it vulnerable to faster depletion.

Figure 3–2 shows regional gas resources and consumption to date. Compare it to the petroleum projection in figure 3–1 for a sense of where the resources are. (The two graphs are comparable; the vertical axes represent equal amounts of energy.)

For both oil and gas there is a clear message: the transition will be much less abrupt and painful if consumption can be held to present levels. The energy experts hope to do it with technology and conservation; I would suggest again that they had better look also at population growth.

Coal

Coal is the most abundant fossil fuel, with recoverable energy about ten times that of the other two together. Comparisons are necessarily crude, since definitions and estimates of "recoverable" coal vary widely. Much of the coal is in seams too deep or thin to exploit, or its heat content is marginal. Bituminous coal is hard on the environment, and lignite or brown coal—about half the total recoverable reserves—is so much worse that it is being phased out even in eastern Europe where it has been a mainstay of home heating and industry. If coal is used, it must be subjected to gasification and other processes to remove the impurities, or the climate effects and pollution will be intolerable (chapter 4).

Meanwhile, for what they are worth, here are the current estimates: worldwide, recoverable reserves are 1.1 trillion tons, 26% of them in the former USSR, 23% in the United States, 11% each in China and West Europe. That total is enough for over 200 years' use at current rates, or still more than 100 years adjusted for consumption rising at EIA's projected rates for 2000–2010. As with the gas projections above, that projection is not adjusted to reflect the expectation that countries will turn to coal as other fuels run out.

For such long periods, estimates of consumption rates become nearly useless. The real conclusion is that there is enough coal for a long time, but the problems of using it without intolerable environmental damage will become central.

■　■　■

For environmental reasons (chapter 4), out of prudence, and in the face of an exhaustible supply, the world must face the transition away from oil and then from all fossil fuels. It is short-sighted policy indeed to wait until the pump runs dry before we prepare for the transition.

It should be a gradual process. It will probably be abrupt. The public anger at President Clinton's fossil fuel tax suggests that this country is not likely to prepare for the change. High European gasoline taxes encourage higher energy efficiency than in the United States, but proposals for further incentives for renewable energy have met a hostile reception. The Arabs are not likely to help smooth the transition. Middle Eastern oil is very cheap to produce, costing well below the world oil price. The Persian Gulf states' self-interest would seem to call for higher monopoly pricing to increase their earnings per barrel and to extend the lifetime of their reserves. However, the Gulf states have lived very high on oil revenues, creating lollipop economies where most of the citizens' needs are supplied free or at subsidized prices. Their governments have become so dependent on oil revenues to support that consumption that they cannot restrain exports to maintain price discipline.

For the short term, conservation can play a larger role in the energy transition than population policy, which works more slowly. Most population growth is occurring in the poor countries, which use relatively little fossil fuel. The Organization for Economic Cooperation and Development (OECD) countries are the principal present energy users, and most of them are close to population stabilization. The United States needs both energy conservation and population policies. However, we will see in chapter 7 that the "emerging" countries, with their combination of population growth and rising energy consumption, are the ones most immediately in need of effective population policies to forestall problems that loom close ahead.

FOOD AND THE ENERGY TRANSITION

While it lasts, there is nothing as cheap as liquid or gas energy gushing from the ground, but that is not an alternative for long. Let us, perhaps reluctantly, brace ourselves for a long secular rise in energy costs.

Since energy is such a major element in the input costs of modern agriculture, a rise in energy prices is bound to lead to an increase in food prices. Agriculture, to be sustainable, is going to have to go through some fundamental changes irrespective of energy costs, but rising energy prices will force some of those changes. They will be beneficial in some ways— but not in raising crop yields to support an expanding population.

Modern agriculture is extraordinarily efficient in terms of labor costs and very inefficient in terms of energy requirements. Crop drying in commercial agriculture, for instance, now uses energy-intensive heaters and fans, but it can be done by sun drying and hand labor. The industrial world may need to reemphasize labor in place of mechanization. The poorer countries may find they should not abandon labor–intensive agriculture. Yields are largely independent of the degree of mechanization. (The principal exception is the need for speed in harvesting, particularly when threatened by rain or frost, or to get another crop in. Labor can be mobilized instead, by releasing it from other jobs for the harvest.) Food raised with more human labor may be more expensive—depending on what happens to energy prices—but the use of more labor would not in itself reduce yields.

That approach has advantages. It would create jobs or save them in an era of widespread unemployment (chapter 5). The use of less energy and more labor would benefit the environment. Organic farming needs a larger labor force than modern mechanized agriculture.

If there is a widespread return to "organic" agriculture, there will be a shift of energy costs but on balance probably a saving. The world will avoid huge investments in chemical fertilizer plants, the generating plants to support them and the costs of transporting the fertilizer, but the on–farm energy costs could well rise. Organic nutrients are less concentrated and more costly to collect and spread than ammonia. Farmers will need more powered machinery or they will have to return on a massive scale to animal power. The latter route has its own costs: the meadows that were converted to grain production or reverted to forests in most of the modern world would again be needed for animals, competing for land that is now much more scarce per capita than it was before the fossil fuel transition.

■   ■   ■

In short, the energy transition will complicate the problems of maintaining food production and will lead to costlier food. The countries already practicing intensive modern agriculture are reaching the stops regardless of energy availability. High energy costs will be a serious block to the expansion of "modern" agriculture in poor countries but less of a roadblock to a sustainable organic agriculture.

NUCLEAR FISSION AND DREAMS OF FUSION

Fission

Whatever the World Energy Council may say, many of the industrial nations realize that dependence on fossil fuels is a transitory phase and that they must find energy elsewhere. They are eyeing a shift to nuclear energy, despite its dangers. It is a known source and an existing technology, and it can even now deliver power at a cost not too much above that of fossil energy. Some 27 countries, including all the major industrial countries, are already using nuclear power, and others will soon join them. Sweden since 1980 has agonized over nuclear energy and still hopes to phase it out, but it provides 30% of the country's electricity, and they can identify no reliable long-term alternative.[41] China, despite its huge coal resources, proposes to expand nuclear power from 2100 to 20,000 megawatts by 2010.

Most societies seem to be well aware of the threat from ionizing radiation, whether from nuclear weapons or nuclear plant accidents such as in Chernobyl. The threat of the use of nuclear weapons by nations or sophisticated terrorists lies always at the back of the mind. We live with the threat, forgetting that it is rising.

There is yet no agreed-upon, safe, long-term storage method for nuclear wastes, yet the prospect is for a boom in nuclear power.

Another problem immediately arises: "conventional" reactors, whatever their design, rely upon exhaustible uranium reserves, which makes them another transitional source of energy. The answer to that problem is a very dangerous one but tantalizingly attractive: build breeder reactors. Start with the spent fuel rods from conventional reactors and reprocess them into plutonium or a plutonium/uranium mix. Produce energy with the plutonium and at the same time generate more plutonium for future fuel. Plutonium is already in abundant supply from existing reactors and from the dismantling of U.S. and Russian nuclear weapons. The breeder reactor is a proposal that the Devil might well have offered Faust.

The breeder reactor is, by common wisdom, exceedingly dangerous and a "dirty" producer of nuclear wastes. The reactor is moderated by liquid sodium at high temperatures and pressures. It is an appallingly tricky process. The French, who rely on nuclear energy for nearly 80% of their electric power and thus have a glut of spent fuel rods, have been trying unsuccessfully for more than 15 years to make their experimental "Superphenix" breeder reactor perform as planned. The Japanese "Monju" reactor is having similar problems.

Nevertheless, nation after nation is quietly deciding to keep its options open. Great Britain, France, Japan and Russia already have plans for plutonium power. Others, such as Switzerland and even Finland, have recently become very hesitant about letting go of spent nuclear fuel, which they could reprocess if it became necessary.

A Rand Corporation study for the Department of Defense concluded that within a decade there will be enough surplus plutonium in the world to make 87,000 "primitive" nuclear bombs. Other authors suggested that the U.S. Government offer to buy up Russian stockpiles to safeguard them and that it press for a worldwide cutoff of the extraction of plutonium from nuclear wastes. The government out of deference to Russia and our allies has decided against the effort.[42]

The dangers from nuclear power are out of proportion to its contribution to total energy production. It provided only 6% of total world energy consumption in 1992, and the Department of Energy expects it to provide only 5% in 2010. It is pursued because of fear of the climatic and environmental effects of fossil fuels, the instability of their supply and above all because of their potential exhaustion. Yet even if you conclude, in light of the energy transition, that nuclear power is necessary, it is very difficult to argue that it is desirable.

Here is the population connection. To take the United States and show what might have been: even with our voracious present appetite for energy, we would not need nuclear power right now—or any imported oil, for that matter—if our population had not risen since the close of World War II. Think of the flexibility we might have had. We would have been in a position to make the transition directly to renewable energy without generating more nuclear waste that we don't know how to store, without opening the Pandora's Box of breeder reactors, and with zero dependence on uncertain and distant oil supplies. It is a powerful argument for a smaller population.

Fusion

For energy, the dream machine is fusion, but it is far from certain that it will ever provide useful energy.

One wise physicist remarked to me that the universe is, in essence, immense energy governed by a set of rules. About the only thing I can understand of Einstein's theories is the literal meaning of $E=MC^2$. We put it on sweatshirts and otherwise invoke it, but we seldom look closely at that supremely simply equation. Energy (in joules) equals mass (in kilograms) times the square of the speed of light (in meters per second). If I have done my rusty algebra correctly, that means that the nuclear energy in one kilogram of mass is 90 quadrillion joules, or the equivalent of burning over two million tons of petroleum.

That perhaps helps us to understand physicists' pursuit of nuclear fusion. The problem, of course, is that it isn't easy. With an immense array of equipment, physicists using the Tokomak test reactor at Princeton in 1994 succeeded in producing a burst of controlled fusion—for about

1/10th of a second, at the unimaginable temperature of 150 million degrees celsius (C).

If they should succeed in producing a sustained surplus, the next question would be, How do we put it to practical work? That problem may never be solved, or the outlines of a solution could be seen in a decade or a year. Nobody knows. Even if a laboratory solution were found, it might take a generation or more to produce usable energy. The conclusion: don't count your chickens yet; don't count on fusion energy to solve looming needs.

Other and more profound questions lie just beyond the practical one. What would happen if we really had boundless energy? Respected physicists like Carlo Rubbia periodically announce they have found a way to inexhaustible power.[43] What if one of them turned out to be right? Would it be a good idea? Is unlimited energy, and the denser population it might support, really desirable? Where would that larger population live, and what would the crowding do to humans and other species? What would the consequences be if cheap energy promoted intensive agriculture and the more indiscriminate use of chemicals?

Professor Albert Bartlett has pointed out that 14 more doublings of energy use by humankind would make the anthropogenic output of energy equivalent to that which we receive from the sun.[44] That calculation is partly a shock treatment. Fourteen doublings is, after all, a 16,384–fold increase. However, it wouldn't take 14. The Earth's ecosystem has evolved in a delicate balance between incoming solar radiation and outgoing radiation from the Earth. Already, human activities threaten to upset that balance by tampering with the "greenhouse effect" (chapter 4). There is some evidence that even a 0.1% variation in solar radiation has a perceptible effect on climate (see note 56). Anthropogenic energy production would match that variation with only four more doublings. Energy use has doubled more than three times in this century alone. Long before we reached that 14th doubling, the additional impact of rising energy use, if energy were readily available, could upset the balance enough to make the Earth uninhabitable by humans and most other species. The simple fact of energy production by any process other than renewables (chapter 4) would eventually pose a limit to human growth, independent of the pollution it generates.

■ ■ ■

Moral: don't trust fission, and don't count on fusion. Even if we had fusion power, there are limits to its use. We cannot grow forever.

Chapter 4

Doomsday Scenarios: Energy, Pollution and Climate

THE EFFECTS OF FOSSIL FUELS have been recognized in widening circles from local "killer fogs" to regional pollution and acid deposition to potential climate effects. The efforts to deal with them provide an interesting lesson in what amelioration can and cannot do.

Conventional energy has immense side effects: local effects such as the acidification of streams and groundwater by coal mine tailings, or terrestrial and marine oil spills; toxic leakage from gasoline storage tanks into aquifers; smog and local pollution; acid precipitation; and climate change. Environmentalists may welcome the prospective energy transition away from fossil fuels, but benign energy sources are a distant and expensive alternative. We must address the demand for energy, not simply try to cobble together ways of mitigating the damage we are doing, and that means reversing population growth or facing some remarkably serious consequences.

KILLER FOGS AND LOCAL POLLUTION

On October 28, 1948, a terrible smog descended on the town of Donora, Pennsylvania, killing 20 old people. (Donora is in the Monongahela Valley near Pittsburgh and was home to a U.S. Steel mill.) That event awoke America to what air pollution could do. It was the symbolic beginning of the American environmental movement. Intensifying "killer fogs" in London similarly galvanized the British.

The first efforts, naturally enough, addressed local pollution. Electric heaters replaced the smoky little coal fireplaces of London, and the chimney pots stopped fouling the air. Control of smoke stack emissions cleared the air of Donora. Most of the industrial world has absorbed that lesson and cut down particulates and local pollution. Still, in New York, Philadelphia, Denver and Los Angeles, the Environmental Protection Agency (EPA) rates the atmosphere as "unhealthful" or worse more than 100 days each year.[45] Even Paris, which is still smitten with the automobile (in both ways), has periodic alerts when old people and those with respiratory problems are urged to stay indoors. In Eastern Europe under Communism, the use of lignite made the air so foul that people in Krakow, Poland, were actually taken down into abandoned mines to escape the surface air, and the statues and gargoyles on ancient buildings dissolved into shapeless blobs in one generation. Those countries are now moving away from lignite. In the third world, reform has yet to happen, and the air in the largest cities is almost unbearable.

REGIONAL POLLUTION AND ACID DEPOSITION

For generations, people have lived with local air pollution and conveniently assumed that it would blow away. One can still see the towering smoke stacks built at power plants in the '60s. They were designed to "punch" the pollution high enough up to disperse it. What they did was spread it regionally. We are learning, very slowly, that there is no "away."

The Fossil Frenzy

The discovery and utilization of fossil fuels, starting with coal, offer a rough parallel to the "nitrogen trap" (chapter 2). The primary offenders are carbon, nitrogen and sulfur. It took millions of years for plants to store all that carbon in the ground and create the atmospheric carbon/oxygen balance in which the human species developed. We have been putting it back into the atmosphere as fast as we can. Carbon dioxide is released in natural processes, but human activities (mostly fossil fuel burning and forest destruction) are adding about 11% to 16% to the rate of natural emissions, and this is sufficient to disturb the natural balance.

Similarly, fossil energy releases nitrogen from automobiles and power plants (aside from the agricultural releases described in chapter 2). Nitrogen of course constitutes most of the atmosphere, but not as oxides. Those anthropogenic oxides return to earth as acid deposition. Some of that deposition winds up leaching nutrients away from forests and fields and into watercourses. Thus it adds to synthetic fertilizers as a source of eutrophication and water pollution.

The same generalization holds for sulfur dioxide (mostly from power plants, with high–sulfur coal the worst offender). We are attacking the natural balance that supports us. In the process, we are polluting the air, promoting global warming and forcing the current rise in sea levels (see p. 63). The human race has embarked upon a dangerous experiment with the one Earth available to us.

Figures 3–1 and 3–2 carry an environmental message: the amount of petroleum and gas still to be used is still much more than we have already used (the amounts below the zero line). Add coal, and the problem is much worse. Resource exhaustion will not solve our climate and pollution problems.

Initial Successes, Present Stalemate

One can limit some emissions, including sulfur and nitrogen, with technological fixes. In the '60s, the United States tended to be in the lead. The Clean Air Act Amendments of 1970 achieved some real gains against both local and regional pollution.

There are multiple sources of pollution: sulfur dioxide, nitrogen oxides, particulates, carbon monoxide, ozone (a necessary filter for ultraviolet rays in the stratosphere but a pollutant at street level) and various volatile organic compounds. These are all by–products of combustion, transformed in some cases by sunlight. As a result of the Clean Air Act Amendments, sulfur dioxide emissions dropped 20% in the '70s and another 10% in the early '80s. Then progress slowed. Nitrogen oxide emissions continued to rise until 1980 but then declined about 6%. Volatile organic compounds declined 17% in the '80s, and some further decline is expected.

The U.S. Government in 1990 adopted the Acid Deposition Control Program intended to cap sulfur dioxide emissions at 63% of their 1980 levels by imposing emission caps on power plants and industry. We have done fairly well compared to most other environmental targets: by 1993 we were down to 84% of 1980 levels. No specific cap was announced for nitrogen oxide emissions, which would have demanded capping the automobile industry, but there again the official statistics show a gain: 82% of 1980 levels in 1993.[46] In a sense, we have showed what technical amelioration can do; and we have also showed its limits.

My initial intent was to put a graph here comparing sulfate (SO_x) and nitrate (NO_x) emissions worldwide, but reliable worldwide data simply do not exist. For what it is worth—and I am frankly dubious—OECD's *Environmental Data* (tables 21.A and 21.B) suggests that the United States generated 52% of OECD countries' total SO_x emissions in 1990 and 53% of NO_x emissions. We are given to crying "mea culpa." A comparison of national data on energy generation and usage suggests that countries' emissions data may not be comparable.

The first wave of pollution reduction required changes such as smoke stack scrubbers, avoidance of high–sulfur coal and catalytic converters on automobiles. Despite fears to the contrary, they add very little to household customers' bills. An EPA study concluded that all costs of environmental cleanup and pollution control have taken about 2% of gross national product (GNP) in recent years.[47] Further progress is going to be a lot more expensive. It will require techniques that strip the pollutants from fuel and permit their recapture. For most pollutants—but not carbon—several such processes have been developed.

The most thoroughly tested process at full scale was perhaps the prototype coal gasification plant at Cool Water, California, with a capacity of 100 megawatts, sponsored by a consortium led by Texaco and Southern California Edison. It is now idle, and I am told that, even though the capital investment is in place and written off, it would not be competitive unless oil and gas were priced at the equivalent of $40 per barrel—more than twice the present market price.[48] And that is before the critical added expense of capturing the carbon dioxide. Some way must indeed be found to make coal less noxious as nations turn to it, but let us not pretend that it will be cheap.

Acid Deposition

After the initial concerns about air pollution and health, another and potentially more ominous issue arose. The growing acidity of precipitation in large industrialized regions seemed to be causing a dieback of trees, the acidification of many lakes and the loss of their fish populations. The danger arose that as soils became acidified their productivity would decline.

An international network of scientists was organized with commendable speed to evaluate those threats. The U.S. component of that network is called National Acid Precipitation Assessment Program (NAPAP).[49] Its subsequent findings have sometimes been controversial. NAPAP was initially chaired by an industrial chemist openly hostile to "alarmism" among environmentalists. Subsequent directors have reestablished its credibility, though—as we will see—many observers regard it as too sanguine. Its conclusions are nevertheless worth examining.

In 1992 NAPAP scientists concluded that the problem is total deposition, not just acidic rain or fog. They discovered that the problem was not simply one of acidity but rather of the synergistic effects of multiple pollutants and stresses. They confirmed that excess sulfur and nitrogen compounds in the environment contribute not only to acidification but also to atmospheric haze, climate warming, low-level ozone and "disturbance to the biogeochemical recycling of other nutrients and metals."

NAPAP discovered hopeful signs: a downturn of sulfate concentrations in precipitation, reflecting the success of the Clean Air Act; a small but less significant decrease in nitrogen concentrations; decreasing acidity of rainfall.

The scientists concluded that threats to forest systems are caused by ozone, acid deposition, sulfur dioxide and nitrogen, in that order. They found no evidence of a general decline in forest health, but high-elevation spruce/fir forests in the East are in decline, and forest soils in the Great Lakes region are being leached of nutrients. They described important new evidence that nitrogen saturation is leaching out nutrients in forests in several regions, but they could offer no overall impact data yet. Fish populations and invertebrate populations in small streams and lakes are affected by "episodic acidification," but the severity and causes are still little understood.

Other observers are less sanguine. An EPA study of the Adirondacks concluded that half the ponds and lakes studied will probably be so spoiled by man-made pollution that they will be virtually devoid of life by 2040. It indicated that nitrogen is emerging as an important part of the problem and that its effects may offset the reductions in sulfur emissions. Conservationist Charles Little challenges the NAPAP conclusions more directly, citing multiple sources to show that forests are in decline all over the United States and blaming it on a whole complex of stresses generated by industrialization, not on acid rain alone.[50]

A detailed study of a New Hampshire experimental forest reports that the forest has nearly stopped growing since 1987 and concludes that acid rain has leached the soil of calcium and magnesium ions that buffered its effects. The weathering of the rocks from which the soil is derived is much too slow a process to repair the damage in the near future, and the success in controlling sulfur emissions is inadequate to arrest the decline.[51]

My sporadic and unscientific observations (of sugar maples in New York and West Virginia, and live oaks in California) incline me to take those critics seriously.

The most frightening possibility of all was raised in 1983 by the President's Acid Rain Review Committee. It pointed out that soil microorganisms are particularly susceptible to a change in acidity and warned that

it is just this bottom part of the biological cycle that is responsible for the recycling of nitrogen and carbon in the food chain. The proper functioning of the denitrifying microbes is a fundamental requirement upon which the entire biosphere depends. The evidence that increased acidity is perturbing populations of microorganisms is scanty, but the prospect of such an occurrence is grave.[52]

This is a remarkably serious warning couched in the understatement of science. It is perhaps the nearest thing to a doomsday warning that has resulted from any environmental problem. NAPAP's scientific oversight board in 1991 warned that NAPAP had not given enough attention to exploring that momentous possibility,[53] and the 1992 report does not indicate that it has yet been seriously explored.

Acid emissions continue, even though the early successes of the Clean Air Act Amendments seem to have ameliorated them. Research has focused on current measurements of the conditions of trees, lakes and airborne pollutants and acidity and, to a lesser extent, on plant growth under acidic laboratory conditions. We do not really know what levels of acidity unmanaged soils may reach, and our knowledge of soil bacteria and their limits of tolerance has not much improved.

Europe, with its denser population and high industrialization, has been the subject of alarming reports about tree damage, but NAPAP cites recent evidence that serious forest damage has been limited to parts of eastern Europe exposed to very high levels of sulfur dioxide pollution from soft-coal burning. Evidence of tree damage elsewhere in Europe is conflicting, but there too the gravest threat seems to be loss of soil nutrients and damage from complex multiple causes, including weakened resistance to pests, pathogens and climate changes.

■ ■ ■

At the least, one can say that the impact of present levels of industrial activity and particularly of fossil fuel use is most unlikely to be benign. Prudence, in turn, argues for reducing those harmful emissions—by better control of fossil fuel burning in the short term and its reduction or elimination as soon as possible—and for limiting the demand that makes the use of fossil fuel necessary.

A QUESTION OF CLIMATE

About twenty years ago, scientists began to realize that carbon dioxide, one of the most innocent-seeming compounds, is the principal villain in the "greenhouse effect." Rising atmospheric concentrations of CO_2, methane, nitrous oxide and other chemicals may be forcing climate warming, which in turn may affect sea levels, forests, agriculture and the human condition. It is worth summarizing what we know and don't know about the effect of human activities on the climate. The climate question is fundamentally important. The answers as to the impact of human activities on climate will largely determine how much human pressure we believe the Earth's ecosystem can take.

As with NO_x and SO_x, numerical data about greenhouse gases tend to grow spongy when one looks closely at the basic data compilations, which are mutually and internally inconsistent. The conventional figure for annual carbon emissions into the atmosphere is about 6 billion tons, but (a) this may be the figure for oil, gas and coal burning only, or (b) it may include cement manufacture, or (c) it may be the total of all emissions, including those from deforestation, biomass burning and the loss of organic matter from farmlands. Deforestation, in turn, may contribute anywhere from 0.4 to 2.8 billion tons and farmlands 0 to 2.0 billion tons.[54] The uncertainty about methane and nitrous oxide is about as great.

This leads to some confusion as to who is responsible. The straight fossil fuel calculation in (a) above shows the industrial nations generating almost exactly half the CO_2 emissions, with the United States alone generating 22.5%. The EIA predicts that the industrial countries' proportion will drop to 46% by 2010. Deforestation, belching cows (!), rice paddies, agricultural wastes and farmland carbon losses also generate greenhouse gases. If these are figured in, the third world shoulders more of the blame. (This is not a popular idea in the third world; see p. 147.)

We do know that there is blame enough for all. The proportion of carbon in the atmosphere creeps relentlessly up. It has risen 30% since the industrial era began and 6% since 1980. That is a very short time. Other "greenhouse gases" are rising too.

Not all climate change is driven by human activity. The great Entrada sandstone strata of our own Southwest testify to eras of extreme dryness inhospitable to most life forms. We don't know where we are in natural climate cycles, but we know that the major cycles are much slower than the human effects. We could reverse a cooling trend or—even more serious—dramatically accelerate a warming one. We have a huge stake in assuring that human activities do not create or drive a warming trend.

The IPCC Climate Projection

Amid a welter of scientific controversy, the Intergovernmental Panel on Climate Change represents the nearest thing to a world scientific consensus on the human effect on climate. The IPCC in December 1995 produced its "Second Assessment" of climate change and its probable results.[55] It is revealing for what it includes and for what it leaves out.

Bear with me. Climate is an abstruse study, and the IPCC reports are not light reading, but they are important. The IPCC scientists have had to come to conclusions about three separate issues: the level of emissions of greenhouse gases, the resultant atmospheric concentrations and the consequences. Then, by their IPCC charter, they have had to suggest a scenario that would avoid serious human forcing of climate change.

Here are their findings, annotated with my parenthetical comments.

- The key finding was that "The balance of evidence . . . suggests a discernible human influence on climate." This is the official answer, after years of debate, to critics who have challenged the scientists to prove human activity is actually causing present world climate warming, as models suggest it should be. (There seems to be general agreement that the past few years have been the hottest on record, except for the brief cooling effect from the Mount Pinatubo volcano in the Philippines, but some scientists argue that it is not necessarily a trend or anthropogenic.[56])

- The Assessment confirmed a rise of between 0.3°C and 0.6°C in global mean surface temperature in the past century and a related rise in global sea level of 10 to 25 centimeters (cm).

- In the absence of policies to reverse the trend, the scientists expect that CO_2 emissions by 2100 will be anywhere from current rates (a remarkably optimistic projection) to six times the present levels, depending on population change and levels of economic activity, with a somewhat narrower range of uncertainty for other greenhouse gases.

- These rates will lead to a temperature rise of 1° to 3.5°C by 2100—faster than any warming trend in the past 10,000 years. The IPCC scientists' best guess is 2.0°C. They state with some confidence that the change will be more pronounced on land than at sea, that the greatest warming will be in high northern latitude winter temperatures and that there will be more winter precipitation in high latitudes. They warn that climate is a complex and nonlinear system and that "surprises" may happen, such as "rapid circulation changes in the North Atlantic" (see p. 99 for more about nonlinearities).

- They project a rise of 15 to 95 cm in average sea level by 2100, with a best guess of 50 cm (20 inches). The changes in both temperature and sea levels will continue in the centuries beyond 2100 even if greenhouse gas concentrations are stabilized at current levels (a formidable task; see p. 69).

The projections are somewhat less dramatic than earlier ones made in 1990 because of a better understanding of the cooling effect of air pollution aerosols such as smog, sulfates and volcanic eruptions. (This is one of the few cases I have encountered where one form of pollution tends to counter the effects of another, rather than intensifying them. The flip side of that statement is that if we reduce the aerosols to control pollution and acid deposition, we contribute to climate warming.)

Let me interject here that this is not the final word. Climate models have been much improved in the past several years, but uncertainties remain:

- the impact of sea ice, vegetation, clouds and humidity on the global solar energy balance. Different types of cloud reflect solar radiation or capture outgoing radiation in different ways. Water vapor is the most important greenhouse gas. The U.S. National Oceanic and Atmospheric Administration (NOAA) has recorded increasing stratospheric water vapor over its Boulder, Colorado, laboratory; it points out that the phenomenon is probably widespread and could hasten the greenhouse effect and also deplete stratospheric ozone.[57] This suggests a fateful synergy: higher temperatures promote evaporation, which increases the amount of vapor, which in turn contributes to the greenhouse effect and still higher temperatures.

- the "disappearance" of much of the CO_2 generated by human activities. The role of ocean plankton and of forests as carbon sinks has yet to be measured with confidence.

- the need, underlined by Assessment Working Group I, for a better understanding of how microorganisms recycle greenhouse gases and aerosols. (Let me emphasize that statement. The soil organisms potentially threatened by acid deposition (see p. 60) may also play an important role in the climate balance.)

- the possible introduction or recognition of new greenhouse gases. For example, the World Resources Institute in February 1995 convened a press conference to warn of the climate effects of "FFCs" (fully fluorinated compounds), by-products of aluminum smelting, semiconductor production, power generation and plastics used in consumer products such as sneakers. They are potent greenhouse gases. They represent less than 1% of the greenhouse emissions, but they are on the rise.

- the possibility of interacting trends, for better or worse. The IPCC study said that little is known about the cumulative effects of multiple environmental and climate stresses such as we are now experiencing.

Despite these uncertainties, the IPCC Assessment is a cautious and sobering warning, reflecting a rising level of scientific confidence in the climate models.

IPCC on the Consequences

The Assessment said this about the effects of climate warming and rising sea levels:

- *Biological diversity.* Some ecosystems may not reach a new equilibrium for several centuries after a climate change, even if the change comes to a stop. This will mean a phase of impoverishment and reduction in the "services" that biodiversity provides to society.

- *Forests.* If atmospheric levels of CO_2 (or equivalent greenhouse gases) double and then stabilize, one-third of the Earth's forests, particularly those in the north, will "undergo major changes in broad vegetation types." In the northern midlatitudes (e.g., the United States), large areas of temperate forests will become scrub or grassland. Even where the shift is from one kind of forest to another, the period of transition will lead to a loss of volume as the old trees die and the new ones establish themselves. This will mean an additional input of carbon into the atmosphere, worsening the greenhouse effect. Some species will die out, unable to adjust to the speed of the climate change.

- *Deserts* are likely to become hotter but not wetter. This will accelerate desertification of arid regions and make recovery more difficult.

- *Aquatic systems.* Biological productivity will increase in higher latitude lakes and streams, but the disruption of wetlands and coastal destruction from rising sea level will have "major negative effects" on freshwater supplies, fisheries and biodiversity.

- *Water resources.* The shrinking of mountain glaciers and loss of snow cover will make stream flow more erratic, affecting hydroelectric generation and agriculture. Reduction of northern permafrost zones will put more carbon and methane into the atmosphere, again intensifying the greenhouse effect.

- *Agriculture.* On balance, no change is seen, as higher carbon dioxide levels promote plant growth and counterbalance the negative effects. This calculation, the authors point out, does not allow for the harmful effect of increased agricultural pests or a likely increase in climate variability. (I would add another consideration: with the drying and warming climate, agriculture will move northward onto acidic boreal forest podzols, which are much less suited to grain production than the present breadbaskets of the world where the soil developed under grasses.)

The regional effects will be more serious. Reduced agricultural output is expected in the tropics and subtropics and in the more arid regions (where hunger is already a serious problem).

The Assessment indicated that climate change and aerosol haze could interact to reduce rainfall at low latitudes even more than climate change alone will do. (Two independent studies have strengthened that hypothesis and suggested that the combination could lead to a 7–14% decline in the monsoon rains on which Indian agriculture depends, plus increased dryness in already arid northwest China. The Indian study assumed a 1.3% per year increase in CO_2 [which is twice the rate used by the IPCC] and a five-fold increase in haze, based on the current increase in haze over India caused by industrialization there and in China.[58] If this model holds up, it suggests even more trouble for the low-latitude third world).

The report does not make clear whether the IPCC factored rising sea level and higher storm surges into its agricultural scenarios.

■ *Human habitat and health.* Bangladesh will lose 17.5% of its land to the rising sea, and Majuro Atoll in the Marshall Islands will lose 80%. Moreover, storm surges now affect about 46 million people annually; the number may double under the "most likely" projection and rise to 118 million under the high projection. (It is unclear to me exactly how this is derived. Population growth alone will generate such an increase without any climate warming, particularly if the trend toward coastal areas continues. By way of example, let me cite the last two Florida "super hurricanes." Andrew in 1992 left a million people without electricity, about 250,000 homeless and damage estimated at about $20 billion—which rocked the U.S. insurance industry (see p. 148). The previous one, in 1947, was said merely to have wiped out "hundreds" of buildings and to have damaged the citrus crop that year—but total U.S. citrus output was higher that year than in 1946. The difference was people—there were six times as many in Florida by 1992—and the loss of much of the Everglades' ability to absorb hurricane shocks because it had been drained for agriculture and housing.[59] People shouldn't live on low coasts exposed to storm surges, and all Florida counties are classified as "coastal.")

Various infectious diseases such as malaria and dengue fever will spread. The vulnerability of different populations will depend on their nutritional status, access to immunizations and population density.

The Problem Narrowly Defined

The predicted consequences are serious enough in their own terms, but I mean no criticism by saying that they are limited by the way the problem has been defined. Because the scientific uncertainties become unmanageable, the authors have not

- taken the analysis of consequences beyond a doubling of atmospheric CO_2 (or its equivalent in other gases), or
- carried the time frame beyond 2100, even though the consequences of even the present levels will extend into following centuries.

Working Group II pointed out that very little attention has been given anywhere to those questions.

Let me again interject a comment here. For most of the potential consequences, the prospect beyond 2100 is completely speculative, but there are dramatic precedents involving sea levels and coastlines. At the end of the Cretaceous, 65 million years ago, carbon dioxide levels were much higher than now, there were no ice caps, and sea levels were much higher. If the Antarctic and Greenland ice caps begin to melt again, sea level will rise far beyond the modest half meter predicted by the IPCC for the next century. A complete melting would raise sea levels about 70 to 100 meters.[60] It would take hundreds of years, perhaps. Some island countries—if there are still countries then—would disappear entirely. So would the Netherlands. Most coastal plains would disappear. They are home to much of the world's population, which would have to move somewhere. The inhabitants of densely populated low countries like Bangladesh would have nowhere to go except to migrate to higher land—anywhere. Sea level in the United States would run along the foothills of the Appalachians, the coastal plain would be submerged and waves would be lapping at the statue on the U.S. Capitol dome.

In 1995, a 1600–square–mile chunk of the Larsen Ice Shelf in Antarctica calved off into the sea, and much of it broke up into icebergs within a week, in an event that "may be unique for this area in the Holocene." The British Antarctic Survey reports a rise in temperature in the Antarctic of some 2.5ºC since 1945. A "dramatic" retreat of sea ice has occurred around the Antarctic Peninsula.[61] Amid controversy, evidence is accumulating that ocean temperatures are warming substantially and that snow cover and sea ice are retreating worldwide.[62] It is both a foretaste of what could happen and itself a potential source of warming. Diminished snow cover means that more incoming solar heat is absorbed instead of being reflected back into space.

Another recent event should warn us that major physical changes to the ice sheets do not necessarily take eons. The USGS in December 1994 reported that the Bering Glacier, a 125-mile-long glacier in Alaska, had "surged" in 1994, moving as much as 300 feet in a day by hydroplaning on water that accumulated under it. There is a theory—still only that—that the great Laurentian ice sheet behaved similarly at the end of the last ice age.

I see nothing in the IPCC report to refute the possibility of a return after some centuries to the climate of the Cretaceous.

What It Would Take to Stop the Engine

The IPCC notes that carbon dioxide, methane and nitrous oxide together cause about 80% of the anthropogenic climate forcing. Most of the IPCC scenarios indicate that atmospheric concentrations (expressed in "CO_2 equivalent") will more than double in the next century. It would take *an immediate 50–70% reduction in CO_2 emissions and further reductions later on to stabilize atmospheric carbon dioxide at current levels* (my emphasis). Atmospheric nitrous oxide has a lifetime of 120 years, and its climate impact is on the way to quadrupling in the next century. An immediate 50% reduction in emissions would be necessary to stabilize atmospheric concentrations at present levels. Atmospheric methane is shorter-lived, and only an 8% reduction would be needed to stabilize it.

Remember (p. 63) that even if the immediate stabilization of those gases were (miraculously) achieved, the present level of climate forcing would continue, and "global mean surface temperature would continue to rise for some centuries and sea level for many centuries."

As we have seen, real world emissions are rising rather than falling, driving atmospheric concentrations further upward.

I fear that it would take a worldwide economic and/or population collapse to stop the atmospheric CO_2 concentration level at a doubling. One rather dyspeptic Finnish commentator in *Science* observed that the IPCC goal of avoiding "dangerous anthropogenic interference with the climate system" is almost certainly unattainable; atmospheric carbon levels have been rising since about 1800 when world population was about 1 billion, before the fossil fuel era. It would be quite an achievement to hold emissions to the current level of about 6 billion tons, to say nothing of returning to pre-industrial levels.[63] The syllogism is perhaps oversimplified, but the warning is valid.

Proposed Solutions: The IPCC Energy Scenarios

The IPCC scientists, with considerable courage, undertook to fulfill their mandate and describe scenarios whereby the human forcing of climate could be avoided—or now (since they have decided it is happening), stopped.

The target: reduce CO_2 emissions by two-thirds to 2 billion tons per year. (As I emphasized earlier, their target would be low enough—but not fast enough—to hold the anthropogenic greenhouse effect at present levels; it would not stop it or avoid the continuing rise in temperatures and sea levels in future centuries.)

They propose multiple actions to achieve the target: use of clean energy, conservation, implementation of new technologies, and a massive reversal of deforestation and land degradation. (Those proposals rest on the unvoiced assumption that societies can mobilize themselves to address climate change when they have not been able to address more immediate problems. The assumption is legitimate—one needs targets—but probably not realistic.)

Here are the dimensions of the problem: they assumed a near doubling of world population to 10.5 billion by 2100, a 7-fold gross domestic product (GDP) growth by 2050 and 25-fold growth by 2100 (including 70-fold growth in third world countries).

I tend to agree with the dyspeptic Finn that they have an impossible job, given those assumptions, and I suspect that even the IPCC scientists may agree. They resorted to "thought experiments" as to how it might be done—in American English, "brainstorming." They came up with five different energy scenarios: biomass-intensive, nuclear-intensive, natural gas-intensive, coal-intensive, and "high demand variant." They concluded that "these exercises indicated the *technical* possibility of reducing annual global emissions" by two-thirds by 2100 (emphasis in the original).

One is inclined to dismiss the idea that worldwide per capita emissions can be reduced to 17% of their present level while per capita economic output is rising 13-fold. That is a 75-fold increase in efficiency. Furthermore, they suggest that "one or more of the variants would plausibly be capable of providing the demanded energy services at estimated costs that are approximately the same as estimated future costs for current conventional energy." Only, I would argue, if they share my assumption that conventional energy by 2100 is likely to be very expensive indeed.

Let us examine their plans for coal. The IPCC scientists would still be getting 17% of world energy in 2100 from coal under the coal-intensive scenario and 40% under the high demand scenario, using more coal than at present in both cases. To minimize the damage, they would use high technology techniques of gasifying the coal and stripping away the pollutants. So far, none of these techniques offers a way of disposing of the CO_2, so they propose to pump it back into the ground, perhaps into abandoned natural gas fields.

This is not as cheap or simple as it sounds. It may be possible, and the IPCC scientists are extremely optimistic about the costs, but caution suggests it would cost nearly as much as it now costs to pump gas out of the ground and ship it to consumers. Carbon dioxide under pressure could resurface. (A corresponding natural upwelling in volcanic lakes in Cameroon suffocated about 2000 villagers in 1984 and 1986.[64]) Add that to the costs of the power itself. Let us take the Cool Water coal gasification plant (p. 59) as a guide. Start with the $40 per barrel cost equivalent already described. Add capital costs and the costs of capturing the CO_2, and one must assume that energy from such a benign power source would cost perhaps four or five times as much as energy costs now.

The problem is both technical (developing reliable clean coal technologies and renewables) and political (persuading governments particularly in the third world to change their ways in their own interest). Right now, most third world countries are installing cheaper conventional boilers, unequipped with the partial emissions controls now mandatory in industrial countries, and devil take the hindmost (chapter 7).

I do not mean to ridicule the IPCC effort. It is a remarkably good one for a consensus document concurred in by 122 governments and 28 organizations. Whether it is brainstorming or daydreaming, the exercise offers many valuable ideas as to the technologies that are needed, the institutional changes to put them into place and the issues of equity between nations and generations that must be solved to do so. Later in this book, I will advocate some of the ideas they propose, such as forcing producers to pay the costs of their pollution ("internalizing externalities," in the jargon) and eliminating subsidies that encourage energy production and use. However, my problem is that they have defined their task in a way that makes it unachievable.

The Competition with Food and Fiber

The question arises, Why not go for renewables? In fact, all the IPCC scenarios have half or more of total energy coming from renewables by 2100. However, their idea of renewables is in large part a move to harvesting renewable biomass—plant materials from annual crops, rangeland and forests. And this is where the energy issue intersects the food and fiber problems of chapter 2. Food is the primal need. I have described the growing competition for land from fiber and industrial uses. Here comes another proposed competitor.

This is the mirror image of the 1980 FAO study that assumed that forest lands could be used to grow food. Both ignore the competition. Everybody assumes that the Earth's resources are available to solve their particular problem. FAO needed the land to "prove" that a growing population can be fed; IPCC needs it to "prove" that a growing population can meet its energy needs without a climate disaster. In real life, food will come first.

Unlike many studies, the IPCC study showed an awareness of the competition, but there is an irresistible inclination to solve one's own problem first, and there is the temptation to stretch every resource to create a scenario that seems to solve the problem. Let me give some examples:

- The IPCC plan calls for converting world forests to a "sink" for carbon (absorbing more than they emit), even as the forests are being cut down and in the face of their own projection that climate change will accelerate the destruction. The forests are not being burned off and cut down simply for exercise. The desperate need for farmland and the worldwide demand for wood drive the destruction.

- It calls for a return to longer intervals for "slash and burn" agriculture, even though population growth is driving peasants toward a shorter cycle (chapter 2).

- The calculations of potential biomass for energy require a remarkable 68% increase in third world arable land between 1985 and 2020. Nothing in recent history suggests that this is possible. Moreover, the effort would require tremendous inroads into tropical forests (chapter 2), which would in turn vitiate the optimistic expectations from forests (above). (The scientists apparently assume that tropical forests make good farmland, which is usually not true.)

- The plan requires 572 million hectares for biomass energy by 2100, roughly equal to total third world croplands in 1985. A doubled population must eat. The plan assumes that growth in cropland, farmed by the intensive methods of modern commercial agriculture, would support both food and energy. I have already expressed my reservations about such agriculture and its prospect.

- The plan assumes a quintupling of fertilizer use in third world agriculture from 1985 to 2020. The impacts of heavy commercial fertilizer use were described in chapter 2; this would compound them.

- It projects production of 10 tons of biomass per hectare by 2050 and 20 tons by 2100. That level of productivity would require heavy fertilization with nitrogen and phosphates, leading to more water and air pollution. A large part of the plan rests on restoring denuded erstwhile tropical forests that have lost their nutrients (chapter 2); this would demand epochal inputs.

- Dung, organic wastes and rangeland grasses would be used for energy production. Too much of this organic material is already being used for cooking fires when it should go into soil building. Another question arises: What would rangeland animals (domestic and wild) eat?

- Most world phosphate resources are in Morocco. Would there be enough? If not, are there other ways of finding phosphates? What problems would arise in developing them? If there are "enough," what would happen to drinking water, groundwater, wetlands, and fishery if we multiply the phosphate loads in the world's water in order to generate "benign" power from biomass?

One could run this exegesis on for pages. The IPCC scientists are simply trying too hard, hoping for too much, and generating problems they do not address. There is a basic imbalance, an instability, about such attempted solutions. Moreover, the effort to solve the immediate problem leads to another question: " . . . and where would we be in 2100, even if all these remarkably difficult remedies worked?" As the ecologists keep warning, you cannot simply do one thing.

It is a lesson that people find very difficult to learn. Would it not be better to extricate ourselves from this whirlpool by looking at the demand side and asking, How many people can be supported without generating this intolerable competition for resources?

Redefining the Problem

The IPCC poses an insuperable task for itself by the way the problem has been defined. Like almost every group addressing almost every problem in the modern world, these scientists fell, perhaps unconsciously, into a fatal error.

They treated population growth as an independent variable to which they must adjust, rather than as a factor that must be controlled if a real solution to their problem is to be found.

Both cause and effect argue for a smaller and less consumptive human population. It would diminish the greenhouse effect, and a smaller population would be in a better position to adjust to the anticipated effects.

The IPCC scientists' solutions evoke an image of Indian villagers running before the Juggernaut, when as scientists and prophets they should be seeking a safe eminence from which to shout to the mindless throng behind the Juggernaut: "Stop pushing!"

The primary cause of the problem is population growth—the number of people to be served. The secondary cause is the level of demand—how many resources each of us uses. The tertiary cause is technological—how we meet that need. The product of these factors is the scale of greenhouse gas emissions. (Here, only slightly disguised, we encounter the familiar formula: I=PCT. Impact equals population times consumption levels times technological choices. We will return to it in chapter 5.) Whether because of their mandate or because of the tunnel vision that afflicts human decision making, the IPCC scientists have chosen to treat the primary cause as a given, to give passing attention to the secondary issue and to focus almost their entire attention on the tertiary variable, technology.

The IPCC scientists recognize the role of population growth in driving the problem. They predicted that greenhouse gas emissions would not rise in the twenty-first century if one assumes very little population and economic growth. (They apparently think that painless efficiencies would reduce emissions enough to counterbalance the limited growth.) The high growth projection, on the other hand, leads to a six-fold increase in emissions (p. 72). Why did it not occur to them to say the obvious? "This entire problem would be reduced to the extent that population growth could be stopped or turned around."

The Cost of Renewable Energy
The IPCC returned, somewhat belatedly, to the environmentalists' pet energy sources: wind, solar, geothermal, perhaps wave power or renewable energy from ocean heat gradients. They don't affect the climate or cause air pollution or acid precipitation. They can be dispersed and local, which saves capital costs and transmission losses. Renewables (mostly hydropower) provide about 8% of current commercial energy use, just ahead of nuclear energy. Most hydropower sites except in Africa are taken, so most of the growth must come in the unconventional renewables.

I would propose that a realistic solution to global warming demands

- less reliance on biomass and a more enthusiastic adoption of those alternative energy sources, and
- a determined effort to hold population growth down to the lowest attainable level.

The problem with renewable energy is that—like truly clean coal—it will probably be much more expensive as the source of base power (the minimal daily power requirement). The wind blows fitfully. The sun shines only at certain hours and sometimes not then. Storage of electric energy is possible but expensive. Sunlight and wind are much more diffuse than fossil fuels. Collecting that energy will involve massive initial investments and a fundamental restructuring of the energy system.

The best practical indicator of possible future cost is the experience so far. Even as supplemental peaking power, which is by far the cheapest way to use it, the cost of direct solar energy is five times that of fossil fuel power; wind energy is nearly four times as expensive.[65]

■　■　■

In trying to be faithful to the scientific caution and tone of the IPCC original, I may not have made clear the remarkable magnitude of what the Second Assessment means:

- At present levels of human activity, we are driving climate toward unknown patterns in future centuries. These will have serious consequences in the next century—bearing most heavily on those parts of the world least able to adjust—and uncertain consequences in centuries beyond, including perhaps a rising ocean flooding coastal plains where much of the world population lives.
- That change will accelerate in all but one of the six projections of future development.
- Even with their best efforts, the IPCC cannot imagine how we could stop the human impact; only with luck could we hold it close to its current levels. Those "best efforts," I have argued, require piling a series of best possible outcomes one on another; they underestimate the other human needs for resources, and the effort would profoundly disturb the Earth's ecosystem if it were attempted.

This really brings us face to face with the problem most people would like to ignore: What level of human population can the Earth sustainably support? It is a question that has been raised many times with widely variant answers.[66] Most of them have assumed a single limiting factor, such as agriculture, without weighing the interconnections that have been described in this book.

Climate warming may provide one approach to seeking a better answer, if it is considered with the other constraints I described in chapters 2 and 3. My Finn's "1 billion" (p. 68) may have been too pessimistic. The calculations of resource use versus technical efficiency, comparing

1800 with 1996, have not been made, so it is not presently possible to make any definitive calculation as to how many people could live on Earth without driving climate change. In the absence of data for a more sophisticated answer, let me take a first cut: one can hope that we have learned enough ways to limit emissions so that, using benign technologies,

a human population of perhaps 2 or 3 billion
could live sustainably at a decent level.

This suggestion is less radical than it sounds. It is where we were two generations ago. Aside from imposing much less demand on energy sources, a population of that size would free enough land to make the intensive use of biomass for energy a realistic possibility.

Even if everybody came to understand the effects of an excessive human presence on the systems that support us and was willing to act on it, the corrections would not be quick or easy. To revert to the UN low population projection: it would take a century before population stopped rising and came back to its present level, even with astonishing declines in fertility starting now. Even with that most optimistic projection, a population of 3 billion is about two centuries away.

The IPCC believes that climate warming will be degrading the Earth's carrying capacity in the meantime. Our target moves away even as we try to approach it.

I have mentioned that there remains enough fossil fuel to wipe out the hope that its exhaustion will save us. Two scientists have calculated (independent of the IPCC studies) that because of the long dwell times of CO_2 in the atmosphere, the rate at which fossil fuels are burned is less important than the total ultimate emissions. They calculated that atmospheric CO_2 will reach 7.6 times the present level shortly after 2200 if fuel and forest burning continue to rise at recent rates. If that rise were immediately stopped, the date would move back several centuries, but eventually the level would be nearly as high. Stopping all forest destruction would only lower the peak to four times the present level. An impossible 25-fold reduction in carbon emissions would be needed to hold atmospheric CO_2 levels to an eventual doubling,[67] which itself is of course far from a solution.

We are left with the conclusion that, if the IPCC scientists are anywhere near to being right, the avoidance of catastrophe depends upon conservation and a transition away from fossil fuels, at an intensity far beyond anything that has yet been entertained by governments, coupled with a tremendous global effort to bring population down to where it can live with an energy system once again reliant upon direct sunlight and renewable resources. Since it is the rich and emerging nations that generate most of this particular problem, the onus is particularly on them.

There is a frontal confrontation here between economic and environmental goals. It is particularly true of third world hopes for economic betterment. Europe is on a demographic path that would (if they wished) permit a reconciliation of economic aspirations and the viability of the planet. The emerging nations are still on the opposite track (chapter 7), and among them China is the only big one within sight of turning it around.

I didn't make those climate calculations.

THE BURDEN OF PROOF

Chapters 1 to 4 have outlined the consequences of population growth and rising consumption. Let me list some of them:

- Human population growth is reaching the stage at which it will be stopped by rising death rates if we do not control fertility.
- The growth is happening very unevenly, creating huge urban slums in the third world, the source of misery and epidemics.
- It has led to the human usurpation of much of the Earth—and we cannot manage it—to the use of energy on a scale disturbing natural balances and to a reliance on chemicals whose impacts on health and the ecosystems we do not understand.
- There is little potential new farmland, and we are degrading the land we have. Populations in drier regions are pressing ever more heavily on water supplies.
- The technology that raised yields to support growing populations is on or near a plateau, and that plateau may turn out to be a peak if practices are not changed to promote more sustainable agriculture.
- Technology has greatly augmented natural flows of nitrogen, sulfates, carbon and phosphates, polluting the atmosphere and water, disturbing wetlands, diminishing fisheries and perhaps degrading the ocean.
- We have destroyed much of the world's forest, with profound consequences for agriculture and climate.
- We rely upon chemicals and monocultures that lead to the multiplication of pests that compete with humans for food, raising the specter of pest outbreaks and famine.
- We are converting livestock and fisheries to adjuncts of cereal production rather than independent sources of food.

- The era of reliance on cheap energy is a very brief period indeed, and yet
- that use of fossil fuels endangers the future of the entire biosystem, including the microorganisms on which it rests, but
- there is no way, if population trends continue, of supplying benign energy at the levels needed, and
- even a draconian population policy could stabilize the human effect on the climate only after considerable damage and perhaps two centuries.

I have left other connections unexplored—industry, mining—but enough is enough.

Let me erect a hurdle here for growth advocates: to argue against a population turnaround, you must prove not just that one of those problems is overstated but that all of them are wrong. There is indeed uncertainty about the precise effects of air pollution and "acid rain" and their interactions, or of greenhouse gases, but very few scientists would dare argue the effects are benign. There is very little doubt about the direction of human impact on land and water supplies or the plateau in cereal yields. Growth advocates have quite a task to justify their faith.

My conclusion may not, by now, astonish the reader: show me one of those problems that would not be more tractable if population growth were to slow, and stop, and better yet turn around. The sooner the better to minimize the damage already in sight.

Perhaps I can drive the point home by inverting it: What goals are served by continued population growth? I can think of only one—the immediate economic stimulus it can provide—and that is a product of bad economic policy rather than a necessary condition (Part III).

CHAPTER 5

Technology: *Deus ex Machina?*

A RECENT L.L. BEAN CATALOGUE advertises a pocket-sized navigation system, downloading data from the satellite-borne Global Positioning System, that can give my location—wherever I am—to within 100 yards, provide bearings and routes to wherever I want to go, store up to 100 locations and correct from true to magnetic north to match my compass. For only $210. I remember using a sextant only 50 years ago to make a (usually wildly inaccurate) star fix at dawn on a rolling deck. It is easy to say "astounding!"

Science and technology are the most exciting frontier of human inquiry today. Beyond their intrinsic fascination, they created the Industrial Revolution and provided the abundant energy that fueled it. They generated the agricultural, public health and medical discoveries that have supported the current spurt of population growth. One can be mesmerized by technology. It has come to be regarded as the provider of all good things and the solution for our problems. No wonder there is a tendency to say "Look what technology can do; it will take care of the population problem."

I will propose a different view.

Don't be taken in by technology. Most of our remarkable new toys are irrelevant to our critical problems, and some technology can make them worse.

Technology has hidden traps. It has gotten away from our political and social ability to manage it, just as scientific methodology has gotten far ahead of the humanities. Technology has made possible a life style unimaginable even for the rich in earlier times, but it is also the leading source of the environmental problems of the industrial world. It has driven humankind into corners we did not anticipate, such as the nitrogen trap. By raising labor productivity, it has helped create a monstrous worldwide problem of unemployment, and it is on the way to making it even worse. We attempt increasingly complex and dangerous "solutions" to each succeeding set of problems. As the complexity has grown, so has the potential for system collapse.

A better solution for each of the problems would be to back off the single-minded pursuit of productivity, to provide enough space to permit more benign practices. This in turn demands that the demand side, population growth and the pursuit of higher consumption, be controlled. We cannot "undiscover" technology, but we cannot simply rely on technology as a ubiquitous solution; we must learn to manage it.

OZONE AND ULTRAVIOLET RADIATION

Perhaps I should start with a straightforward connection. We are assaulted by threats we did not know existed a century ago. The level of ultraviolet (UV) bombardment of the Earth has been rising. The cause is the destruction of stratospheric ozone, largely because of the use of chlorofluorocarbons (CFCs) in air conditioners. Stratospheric ozone levels fluctuate. Over Antarctica it is seasonally depleted by as much as 70%, over Siberia in late 1994 by 35%. Over the north hemisphere midlatitudes it is declining by about 4% per decade. A NOAA scientist described total UV radiation as about 30% above pre-ozone depletion levels.[68]

We are taking refuge in our air-conditioned cars and buildings to avoid a growing source of skin cancer and eye cataracts. Other forms of life do not have that refuge. Field experiments with ocean plankton and with amphibians show that UV radiation can damage eggs and affect fertility. This is a cause of the worldwide decline of amphibians.[69] It would be sad if the pursuit of human comfort led to the wiping out of genera that have survived earlier natural waves of extinction. It will be disastrous if we are the agent of destruction of ocean plankton (see the discussion of fisheries in chapter 2).

There is good news and bad news. There are substitutes for CFCs. The industrial world is phasing out their use under the Montreal Protocol of 1987 and its annexes. This is an avoidable form of pollution and one of the few that is not tightly linked with human numbers. The bad news is that the best substitutes are HCFCs (hydrochlorofluorocarbons). True, they don't attack stratospheric ozone, but they are a serious greenhouse gas, and the IPCC projects their production to rise from negligible amounts in 1985 to 1.6 million tons by 2025 if CFCs are phased out, which is 2.6 times the 1985 tonnage of CFCs. Solutions create problems. We shift to a greenhouse gas to run our air conditioners in order to avoid ozone depletion, and we make the climate hotter, so we need more air conditioning, so we make the climate even hotter. And so on . . .

CFCs and stratospheric ozone offer another lesson. The time sequence is revealing. Freon (CFC) was introduced as a refrigerant in the 1930s, and it was nearly 40 years before Mario Molina and Sherwood Rowland discovered what it was doing to stratospheric ozone. That was long enough to create the ozone problem. The point here is that the benefits are quickly appreciated, but the damaging consequences may not be recognized for decades or generations. We can perhaps find another substitute for CFCs or if necessary make do with less air conditioning (though that may not be a popular prospect in the face of global warming). The cost of backing away from other innovations that turn sour may be much higher, especially if population has grown and depends on the innovations (e.g., the Irish and the potato) and if no benign substitutes are in sight. Population growth thus makes it much more painful to give up activities or discoveries that prove dangerous—and figure 1-4 reminded us how little we really know about what might prove dangerous.

Methyl bromide raises another aspect of the problem. It destroys stratospheric ozone even more effectively than CFCs. The pesticide is being phased out in the industrial world but is used in the third world on grain and tobacco crops. At a recent meeting of Africans, sponsored by the UN Environmental Programme, the participants made clear they had no intention of phasing out its use until they had a safe and effective substitute, because they cannot afford the crop losses in the face of the needs of expanding populations.[70] There, in a nutshell, is the dilemma that population growth imposes on environmental policy.

TECHNOLOGY AND WASTE

It is easier to think of the savings from technology—insulation, the telephone—than of the waste it generates. Perhaps the most difficult of environmental mind-sets to achieve is the recognition that most human activities generate waste—more waste, in fact, than economic products (by volume). And technology has been a leading cause.

Carbon, nitrogen, phosphates and sulfur (chapters 2–4) are not the only things technology sets loose in the environment. The industrial emissions of lead into the atmosphere in 1990 were 12 times the natural releases. For copper, the ratio was nearly 6 to 1, for cadmium 5 to 1, for arsenic

1.6 to 1.[71] As with our chemical releases, we don't know all the consequences, but with some of them the immediate harm is well documented.

Industrial Processes

Industrial processes regularly concentrate natural materials, making them more usable but also making them and their waste by-products more toxic. The exploitation of petroleum has generated a range of issues: oil spills, polluting refineries, the buried gasoline tanks that serve the automobile and leak into aquifers, climate change and the insults to the atmosphere from all those automobiles. Vegetable matter decomposed gradually into crude oil without much effect on the biosphere. Humans created a disturbance by extracting and using the crude oil. Uranium in nature is relatively benign. The decision to use it for explosives and power generation has led to a horror story. In two brief generations, we have created wastes that will threaten our descendants' health for millennia. To contain them, we must master the unprecedented feat of predicting the geological state of the repositories and assuring the physical security of the sites hundreds of thousands of years from now.[72]

We need a period of consolidation, not driven by rising demand, to learn more about the consequences of economic activity, and we need a foresight process (chapter 18) to organize our ability to weigh the consequences of our actions.

The Automobile and the Atmosphere

Let us look at that favorite whipping boy, the automobile.

We think of movement—freedom—when we think of the automobile. Let us rethink the image. Its primary product is pollution. Carrying just

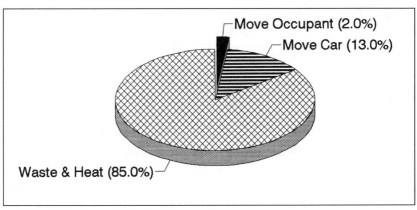

Fig. 5-1. Auto Inefficiency – Ford Escort; single occupant
(Science, *August 10, 1984, p. 591*)

the driver (as usual in the United States), its useful product is just 2% of the energy it uses—and that says nothing about the energy that went into its manufacture, into the highways that carry it or into the sprawling development that it makes possible. Most of the energy is dissipated as useless heat or air pollution. In the United States alone, trucks and automobiles generate some 8 million tons of nitrogen oxides, impeding any effort to control atmospheric pollution. It is a convenient but remarkably energy-inefficient technology compared to the traditional pedestrian or bicyclist. If you like the automobile, join me in arguing for fewer people, because the Earth can take only a certain amount of such an inefficient "solution."

Technology got us into the problem, and technology may help get us out. The electric battery–powered automobile is being touted as the solution, but it would increase the demand for electricity and contribute to more sulfur dioxide pollution even as it created less nitrogen pollution. A proper respect for the environment would lead us in another direction: the hydrogen fuel cell. It is probably not far over the horizon.

Daimler-Benz has a fuel cell–powered experimental minivan on the road, operating on methanol rather than directly on hydrogen.[73] Even using methanol, it is less polluting than the internal combustion engine or the electric battery car. It also provides a way through the energy transition with minimal disturbance. Since gasoline tanks can hold methanol, the intermediate phase of redesigning service stations into battery-charging stations can be avoided. When the technology is ready—and if water can be economically ionized into its hydrogen and oxygen components to power the fuel cells—those motors can readily be converted to pure hydrogen, which is nonpolluting and does not contribute to global warming. Sitting above a pressurized hydrogen tank is a bit daunting, but we live with natural gas and propane tanks already. Gasoline itself is far from benign, whether it is ignited or allowed to evaporate or leak into water tables.

A partial solution to the environmental impact of the automobile, perhaps, but not necessarily a sufficient one (p. 94).

TECHNOLOGY AND UNEMPLOYMENT

Technology is anything but a solution to the modern problem of unemployment. It is, along with population growth, the principal source of the problem.

Technological Displacement

There are two faces to technology, and both are aspects of its success in raising labor productivity. It offers the dream of a world with enough for all, without masters or slaves. On the other hand, it is an intensifying threat to employment. Technology has increased productivity to a point where the overpowering issue now is jobs for the unemployed, not higher productivity for those in the system.

The writer of a recent *Wall Street Journal* article pointed out that there is something new about the new computer, electronic and communications technologies: " ... tractors put only farmers out of work, and machine tool automation only factory workers, but smart devices and computer networks can invade almost every job category ... No technology has ever been as protean, so capable of cutting huge swaths through unrelated industries ... "

The writer listed a frightening array of new programs and technologies and the jobs they are affecting or eliminating. He took note of the traditional economic wisdom that those eliminated will find other and better jobs, but he observed that a growing number of economists are beginning to question that wisdom. He described the U.S. experience of southern farmers displaced by the tractor, who went north to factory jobs that were lost to automation, and finally to joblessness and a role in the "devastation of the inner cities"—and those were the less devastating technological innovations of several decades ago.

The same article includes a quote from Robert White, then president of the National Academy of Engineering: "The creation of new industries may not provide enough jobs fast enough to replace those lost ... "[74]

Productivity generates problems that are not being addressed. Operating a backhoe looks like fun, and there is no joy in digging a ditch with a shovel, but what happens to the shovel wielders who are replaced by backhoes? We are far beyond the stage where we can toss the question off with the assertion that they can go on to more productive and interesting work. Nations must find a way of integrating those consequences of technological change into their development process and not simply adopt every new technology because it is cheaper.

Technological displacement of labor has been around at least since the invention of the wheel, and it is a transitional price that humankind has paid for higher productivity and a better life. The problem right now is that it is happening very fast and it is paired with unprecedented population growth.

Technological displacement now affects all income levels, but perhaps its most frightening aspect is that it undercuts the less skilled, who constitute most of the extraordinary population growth in the third world and who are (as we shall see) the most deeply afflicted population group in the industrial countries. It is so efficient that it can displace even very low income workers, and the reach of communications is such that it is invading even low wage countries and further aggravating their desperation.

The American economist and Nobel laureate Wassily Leontief years ago predicted the labor effects of the technological revolution. He observed that with very high productivity there are limits as to how many goods and services a society can absorb. (I would add "or the environment.") The result is a small and prosperous technological elite and a vast number of marginal service workers and jobless. He has suggested that modern societies will either have to limit work hours or arrange for massive income transfers to those not in the system.[75] (I would add that the first proposal would be circumvented as workers moonlight to occupy their idle time. The second requires an expansion of welfare, which poor countries cannot afford and this rich country, at least, does not want.)

Leontief thus undercut the simplistic but often quoted argument that every person has "one mouth, but two hands to produce with."

The need for a role in society is perhaps second only to the elemental needs for food and shelter as a human requirement, and in modern society a job is the way that most people find that role.

The International Labor Organization (ILO) in its 1994 World Employment Report said 30% of the world labor force is unemployed (including 35 million in the industrial countries) and described the situation as the worst since the Great Depression.[76] Unemployment is apparently even more prevalent than hunger in the world right now, which makes it the single most immediate human problem here on Earth.

Societies must find a way to provide a role with honor for those who cannot handle the technological sophistication of the modern labor market. They had a place in the system. They don't now. The feeling of irrelevancy, the loss of self-respect, must be a terrible source of despair, self-hatred and alienation and may be a source of violence. Helping such people to find a niche must be a part of the foreign trade, labor, welfare, taxation and even scientific policies of any humane nation. That thought will be a leitmotif throughout this book.

To live in this hypermodern world, prospective workers need the best possible training in order to function. As we will see, both poor countries and the United States have more young people than they can train to the high standards of a technological society. But rising productivity means that there would not be enough high-tech jobs even if they were trained. A population policy would ameliorate the problem for poor and rich countries alike, though it alone is not a sufficient solution (chapter 12).

Complexity and Human Skills

Can everybody handle the level of technology demanded of them? Do the skill levels of the jobs that technology has created "fit" the abilities of those who must try to master them? Is everybody infinitely trainable to handle jobs that grow more complex each year? The questions are unresolved. There is some hope. People drive cars, usually without disastrous results, despite dire early predictions of the carnage that would happen if everybody were given control over such complex and dangerous machines. On the other hand, most of us have watched the deterioration of services as unqualified clerks try to run computer programs that are over their heads. Consider the mess in Washington, D.C., where the government accounts have degenerated into such total chaos that two audit firms gave up and said that it was impossible to audit them. It is exciting to challenge people to new heights, but we may be pushing the skill requirements above the limits of the people who are asked to perform the tasks. With our present educational failures, that is a certainty, not a speculation. We must look honestly at the question whether theoretical efficiency achieved with more and more complex systems is mismatched with the needs of people for work they can handle.

■ ■ ■

There is, as usual, a population connection. The least educated are the most fertile. I will suggest (chapter 13) that the "two-child family"—as the socially condoned upper level of fertility for everybody—would help to match the numbers of those seeking work at various skill levels to the demands of a technological age. There is, however, a delay of about 20 years between the time that fertility starts down and the time the decline shows up on the labor market. The other benefit is more immediate: lower fertility would permit higher per capita expenditures on better education, which—not incidentally—would probably encourage still lower fertility and thus help to arrest population growth.

COMPLEXITY AND INTERDEPENDENCE

In the 1920s, E. M. Forster wrote a memorable and ominous short story called *The Machine Stops*. It described a highly efficient society in which all wants, including air circulation, were provided by a vast and complicated machine. Today one would call it a mainframe computer. The story, as the title suggests, is about the inhabitants' utter helplessness when suddenly the hum of the machine stops. A grim little fable and perhaps a prophetic warning of the dangers of too high a degree of complexity and interdependence.

Modern economies are built on both trends. That is what the Industrial Revolution and now the "information revolution" have been about. Most people don't seem to see this as a problem. Economists usually dismiss the possibility of "surprises." The American middle class, to judge by its low rate of savings, agrees. Some of us who remember the Great Depression, when the machine stopped, still worry about surprises. To anybody who does, flexibility of a system is as important a goal as productivity.

Modern societies can absorb shocks better than poor ones, by and large, but that is a function of their wealth, not of their complexity, and of their lesser dependence on imported food (chapter 7).

Few people would really want to go back to an age so simple that they had to draw their own water or chop their own fuel supply. I don't plan to give up the wheel. I love the computer on which I am writing this book. The rejection of technology will get us nowhere, but again, I urge prudence.

The Shrinking World

Two things in particular have shifted the economic relations among nations: technology and cheap fuel. From Samuel Morse and the telegraph (1844), through Marconi's successful radio transmission across the Atlantic (1901), we have progressed to the semiconductor, the modem and the ability to manage and transmit tremendous amounts of information, and to a world made accessible via satellite. There are American corporations having their daily books processed in India because accountants can be hired there so cheaply.

The communications revolution has been matched by the ability to move goods cheaply and swiftly. The annual movement of 200 plus million tons of grains, 400 million tons of coal and 2.5 billion tons of petroleum in international trade would have been unimaginable in the days of sailing ships. Add to this the ability to move people and valuable cargo by air and you have the technical basis for global commerce.

Interdependence versus Flexibility

Generally, people celebrate new technologies, but our institutions have not had time to adjust to them. Technological interdependence, expressed in international trade, promotes efficiency and, in theory, better living, but that efficiency itself has led to unemployment and declining income in both the developed and the poor countries.

With luck, interdependence may even discourage future aggression by making the costs of disrupting world systems too high to contemplate, but the flip side of that argument is greater insecurity (chapter 8). Can we really hope that such a vast machine will stay operable? The very interdependence that makes disruption of world trade costly can become a liability when it is disrupted. Instead of discouraging aggression, it could make it inevitable. I have cited the example of U.S. oil dependency on the Persian Gulf (chapter 3). A similar crisis could occur if there were a breakdown in the world trade pattern for food, for Moroccan phosphate fertilizer or perhaps for other critical items where the supply is highly concentrated, such as industrial spare parts. That contingency is an argument for incorporating flexibility into national policies.

Domestic systems can also be vulnerable to complex and integrated systems. I have an early memory of half–built rural power lines stopped in the middle of nowhere in the Great Depression. It wasn't a catastrophe then because the farmers had not become dependent on electricity. It would be a catastrophe now. The resilience and flexibility of the economic systems we build are perhaps as important as their efficiency. We have pursued efficiency at the expense of flexibility, because rising populations need goods, and technological innovation and expanding markets made them cheap. Cheap is not everything.

■ ■ ■

To be specific as to the remedies: quite aside from the other arguments for renewable energy, there should be a premium on developing dispersed (and renewable) systems. For energy, this is an argument for flexibility and local energy sources in the power net, even if they are somewhat more expensive than conventional fossil fuel power plants. There should be a premium on a smaller population if only as a way of avoiding dependence on integrated worldwide systems to supply the food, energy and other needs that are central to human well–being.

The Narrowing Effects of the Technological Revolution

Complexity has a consequence that seems to draw very little comment. The explosion of data is heralded as leading to a new "information age." I suspect it may lead to a narrowing rather than a broadening of intellectual horizons. The Internet myth is riding high right now. Supposedly, it will make everybody with a computer into a universal genius. Yes, it makes research easier. No, it does not open vast new horizons. The human mind is a remarkable computer but a relatively small one. How much more information can it handle?

Academia has responded to the explosion of knowledge by carving out progressively narrower areas in which a specialist can claim to keep up with the literature. We are all moving in that direction. An environmentalist can keep up with the edge of knowledge concerning perhaps one area of environmental issues, such as climate or radioactivity or waste management. If an environmentalist cannot hope to encompass all environmental issues, how much less can a politician, who must juggle dozens, hundreds or perhaps thousands of other issues equally complex—starting with the task of getting elected. They become dependent on experts, who may disagree among themselves, and who in turn will not understand the relationship of their specialty to other issues.

It becomes progressively harder to know enough of different disciplines to relate them to each other, and yet it is precisely this capability that must save us from the accidental by-products of our actions. This alone makes the systematic development of foresight machinery (chapter 18) terribly important because it provides a way of countering specialization by drawing upon different disciplines. Even so, no institution is a substitute for the ability to bring the disparate threads together in one mind.

Change has been remarkably fast, and we are at an edge where there are clear signs that neither workers nor scholars nor political leaders can handle the technology and breadth of issues they are expected to manage. *Seek simplicity.*

TECHNOLOGY AS DESTROYER

To continue the indictment: war has become more deadly and has reached into civilian populations as a result of technology. Nuclear weapons are the culmination of a process that has been going on for most of history but, like other processes, has speeded up dramatically.

It makes for a more vulnerable world, even as nations become more interdependent. Iraq was developing radiological and biological weapons—anthrax, botulism and aflatoxins—banned though they are by international treaty, when it miscalculated and invaded Kuwait.[77] Nuclear, chemical and biological weapons are now within the reach, not just of nations, but of sophisticated criminal organizations ambitious enough to disregard the very apparent dangers to themselves. The Pentagon has launched a billion dollar program to develop sensing devices and counter weapons.[78]

Mankind's capacity for irresponsible disruption has come a long way. From throwing rocks to fabricating "the bomb," and now releasing pathogens. In 1995 a fanatical group released a deadly aerosol into a crowded Japanese subway. An Aryan Nation white supremacist was arrested in Ohio after he had gotten his hands on bubonic plague bacteria by posing as a bacteriologist. An Arkansas farmer was caught with stocks of the deadly poison ricin.[79]

Automation has lowered the accident rate in modern industry, but accidents become progressively more serious as technology releases more potent forces. When wood was our primary source of energy, the loss of a woodlot might threaten the owner's comfort but nobody else's—and there was more wood. An explosion in a steam-generating plant might kill the few people in it. The meltdown at Chernobyl forced the evacuation of 100,000 people.

■ ■ ■

This aspect of technology would not be relevant to population growth except that, in raising the competition for space and resources in an already crowded world, population growth multiplies the tensions that can lead to technologically sophisticated terrorism. For one example, see pp. 150–154.

THE USES OF TECHNOLOGY

There is still plenty of useful work for science and technology if there is a sense of direction. I will leave to the scientists the notoriously tricky task of trying to identify the directions in which science will move, but I can offer some fairly obvious thoughts as to what would be useful:

- The problem now is a finite Earth bearing an unprecedented load. One area in which the benefits are unarguable is benign technology—more efficient processes that satisfy human wants with less wasteful use of energy and resources. Such techniques exist particularly in energy generation and use.

- Low-till agriculture saves energy. Perhaps genetic engineering will substitute biological processes for industrial energy in converting nitrogen. Better storage techniques increase the effective harvest at little cost.

- Technology can help to fight the resurgent agricultural pests and human pathogens that are becoming resistant to the pesticides and medicines that technology itself has thrown at them in the past 50 years.

- Science may develop salt-tolerant crops that can be grown on lands now abandoned or underproductive.

- Science can help to control the pollution that technology has generated. "Biochemical remediation" of toxins such as PCBs is an example, and promising techniques are being explored. Cattail swamps are being used to process sewage effluent, and mustard plants are being used to take up heavy metals such as selenium. The residents of Cape Cod are trying to save town water supplies by using a filter of iron filings to neutralize the solvents, jet fuel and some very dangerous chlorine compounds that the groundwater is carrying from a nearby military base. Microbes can clean up oil spills such as the *Exxon Valdez* created.[80]

- Fuel-cell engines for automobiles and small machinery are on the way. They will save energy and reduce pollution (p. 82).

- Thin-film solar cells are at the exploratory production stage.[81] They may well be the ultimate form of renewable energy.

- We still need effective and cheap contraceptives, including an effective "day-after" pill. This area of research has been largely abandoned, particularly in the United States, because chemical companies fear retaliation from "right-to-lifers."

Having said all that, let me emphasize that most of the problems we now face are not technical ones. Technology is no substitute for tackling population growth. We must learn to evaluate the environmental impacts of technology much more thoroughly than we have been willing to do.

Until the problem of unemployment is mastered, the world is ill served by new technologies that further increase productivity. Unfortunately, technology right now is going in exactly the wrong direction, by developing smart machines, sophisticated analytical programs and networks that are throwing the skilled as well as the unskilled out of work.

It will be hard to stop that pattern. Societies have never attempted systematic management of the direction technology takes. They act after the fact by regulating or limiting clear and present dangers such as unsafe aircraft or (sometimes) insufficiently tested drugs. The resistance to being told not to make use of an efficient technology would be enormous. I will be called a "Luddite" for making the suggestion. Luddites were desperate unemployed workers in early nineteenth century England who took to smashing the new machines to try to save their jobs. Conventional wisdom since then has excoriated them for trying to block progress. Perhaps, but the combination of technical change and the Enclosure Movement (p. 138) was making early industrial England a very difficult place for workers. At least they had a New World to escape to, and displacement was less pervasive than now.

■　■　■

Very well. If we cannot manage technology, we must face the consequences. Better than that, we should limit them. If we cannot keep technology from destroying jobs, we must limit the growth in demand for jobs.

CHAPTER 6

A Conserving State of Mind

HUMANKIND RIGHT NOW NEEDS a coherent vision of where we are headed and a way of considering the consequences of technological change, much more than it needs new technology. We need a state of mind that understands the role of the human species on Earth and that seeks to identify the consequences of our activity, an attitude that is given more to reflection than to impetuosity and that puts foresight among the highest virtues.

Conservation is not always simple. It requires a new vision of economic activity that integrates it with the issues I have been describing—asking of every economic process, What is the overall cost of this way of doing things, including the social and environmental costs? Could another way of doing it bring those costs down to tolerable levels? Would a smaller population help us get there?

We need to re-examine our social customs and behavior. Something is indeed out of whack. We light up our cities at night for fear of our own species and we thereby destroy the embracing darkness and waste much of our energy on a very questionable amenity. The International Dark Sky Association is trying to tell us that we lose a sense of the night and darkness and gain very little from our wasteful lighting practices. Crime is higher in those brilliant cities than in the dark countryside, and they have lost something else. A friend once told me of a Black inner city teenager who, taken to a summer camp, looked up at the stars and asked, "What are those little white things up there?"

Americans wear three-piece suits or the equivalent in our semitropical summers, seal our buildings to keep out the noise, the dust and the larcenous, and then air condition them to make life bearable—and the energy we use helps to make the climate hotter. I started my career in the old Marine Guard Barracks amid the trees on the Ellipse in front of the White House. We opened windows and went coatless in the summer, and it was better.

It is this point that brings conservation and population policy together. The Juggernaut is not just people; it is a species that consumes at a high rate, often without knowing quite why.

CONSERVATION AND FORESIGHT

Let me use architecture for both bad and good examples. The skyscraper era culminated in the International Style—big glass boxes like the Lever or Seagram buildings in New York—that flourished from World War II until the 1960s. They are bold and dramatic and monuments to human arrogance. The builders had money and cheap energy, and they were making a Statement, so the buildings were built as sculpture. Much of the core of the building was given over to elevators, making the design inefficient and energy-intensive. The architects apparently never considered the path of the Sun or what it would do to their edifices. There was little insulation, and the big windows were the same all the way around. On the south side, people sweltered, while on the north side they froze. Those on the east side were blinded in the morning and on the west side in the afternoon. The buildings after 40 years are still leaking energy and soaking up heroic levels of air conditioning and forcing their unhappy occupants to rig sunshades and to bring in portable fans and heaters. Until they are torn down or rebuilt they will contribute to air pollution, energy dissipation, intensified ultraviolet radiation and global warming.

Solar construction is the antithesis: orientation to work with the sun rather than fight it, sunbreaks and windows to deflect the summer sun and invite the winter sun in, masses to hold the solar heat, smaller windows and better insulation on the north, shade trees and smaller buildings with fewer elevators. Those buildings require a smaller urban scale, which means fewer huge cities; and that can work, worldwide, only with fewer people.

Skyscrapers are still being built—even as their economics are coming under scrutiny. Solar architecture is still an oddity. We learn slowly, here in the shadow of the energy transition. We need some better organizing principle for making decisions.

Automobiles are another good case study because they are so central to many environmental problems. The fuel cell engine is a legitimate technical "solution" to the problems of the internal combustion engine since it would help to reduce air pollution, acid precipitation and climate change (p. 82). Incremental technical solutions have their place, but they are not a substitute for a broader conceptual approach to conservation. Without a much clearer view of where the automobile is taking us, people are unlikely to give up their cars, or dreams of them, unless they have to—however much environmentalists may berate them.

If we in the United States had had planners and they had looked ahead, we might have kept streetcars and light interurban railroads rather than subsidizing roads and the automobile. Nations just starting on the industrial path could still do it, but they probably won't (chapter 7). Instead they will follow the older rich countries and, when the pollution becomes intolerable, then look about for palliatives that may well generate other problems, just as the United States is doing now. There is a certain squirrel-cage quality to all this, and a fluctuating risk as people hope for new technologies to save us from the successive problems we generate.

There are ways to look ahead. Those industrializing societies could judge whether a technical fix is enough or whether a different fundamental direction is more beneficial. That process of assembling the potential consequences is called foresight. After they looked at all the costs and advantages, they might well conclude that different transportation and settlement patterns would be a better choice. Back, perhaps, to light rail.

Foresight thus becomes an integral part of conservation, along with a population policy. I will come back to this argument in chapter 18.

The standard retort is that humans cannot see consequences clearly enough to make such farsighted decisions. My response to that objection is that it is wiser to try than to resign one's society to mindless change.

WASTE AS AN ASSET

If waste is our primary product, perhaps we should rethink it. It is often a useful product that is simply in the wrong place.

Industry and Energy Wastes

Most of the chemicals we produce wind up as waste, dumped indiscriminately in landfills. The industrial nations have begun to pass legislation regarding their safe disposal, but we often have only a hazy idea of what "safe" is. The new industry of illicit dumping, on land or sea, demonstrates the resistance to such mandates. Apparently, many people are committed to business as usual, even if it kills us. When industry is small enough, the environment can buffer and absorb its wastes. As it grows, the environmental impact becomes intolerable, which leads to regulation and in turn to the industrialists' complaint that their freedom to operate is being taken away.

Our present economic practices involve little or no charge for throwing most materials away. Society is justified in making the polluter pay the cost of pollution. (This "polluter pays principle" is enshrined in a 1974 OECD decision that is generally ignored.) If producers had to pay the real costs, including "externalities" like pollution, it would be a different ball game. Sulfur, nitrogen and carbon dioxide play a dual role: they are fertilizers and industrial inputs as well as pollutants. Power plants would seek markets for such "wastes" if they had to pay the social costs of discarding them. That alone would be sufficient to make many such beneficial synergies profitable. We need to begin thinking in those terms and changing our haphazard system of taxes and subsidies accordingly.

Sewage

The scale at which all wastes are produced is related to the size of the population to be served, but the connection between sewage volume and human population is singularly inelastic.

Billions of dollars are spent to manufacture fertilizer, and eventually we flush many of the nutrients down the toilet. We may not be able to continue that modern custom. Repugnant as the thought is to fastidious moderns, perhaps the Chinese were right: put those nutrients on the fields, not into water sources. As population grows, this sort of rethinking may become increasingly necessary to enable us to avoid overwhelming our planet with our wastes. If we cannot reduce the population that creates these artificial flows of nutrients, we can recycle them rather than using them on a one-time basis and turning them loose to upset natural systems. The industrialized countries have begun in a small way to use sewage plant sludge as a fertilizer on parks and golf courses, but we are capturing only a tiny fraction of the nutrients that otherwise head downstream. To use them on crops does not pose a problem of microbial infection (contrary to popular belief), but we must find a way to get rid of the chemicals and heavy metals that get into sewage sludge.

■　■　■

This itself would be a stopgap and a partial solution; the nitrogen would still find its way into water and the atmosphere, with dire results (chapter 2). Nothing really replaces a population policy.

BIODIVERSITY: DESTROYING OUR FRIENDS

The Earth looks so vast that it is hard to remember what a remarkably thin zone is livable. The great majority of Earth's organisms live in the upper 50 meters or so of the ocean where sunlight can penetrate, or on land in a fragile veneer stretching from the treetops to a foot or two below the ground. For dramatic evidence of how thin the biosphere is, look at a highway road cut. A gang plow in a field literally overturns most of the biosphere in a 20-foot stripe as it progresses. The life that surrounds us is far from infinite or indestructible.

Vaclav Smil nominated the process of synthesizing nitrogen fertilizer as the most important innovation of the past century (p. 22). Perhaps I can nominate another surprising candidate: the modern circular-bit drilling rig. A century and more ago, when John Henry was running his race with the steam-driven boring machine, humans' ability to do damage was limited by the sheer work involved. We have that ability now, perhaps to our regret. We have an almost unlimited capability to extract sulphur, phosphates, carbon and metals from the Earth and the air and spew them into that narrow and fragile zone that contains the biosphere.

I am astounded by the human indifference to the consequences of our activities. It could prove suicidal. We modify the atmosphere we breathe, with remarkable sangfroid. We destroy the plant and animal species with which we have coexisted and on which we may depend. A recent UN estimate of the rate of destruction of other species by mankind was 1000 times the rate of disappearance before our arrival. Our impact may match the Cretaceous extinctions 65 million years ago.

When we think of other species, we tend to think of the big, visible ones. The ones we cannot see may be the crucial ones. Each cubic meter of earth, I am told, is occupied by literally billions of such organisms—and humans are usually unaware of them and of their critical role. Research has yet to dismiss the warning of the President's Acid Rain Review Committee (p. 60). Human activities could make the Earth uninhabitable by destroying the ability of earth microorganisms to stabilize the system. Unlikely—one may hope—but by no means impossible. As Professor Lynn Margulis is fond of pointing out (to the intense annoyance of traditionalists), microbes are the most important part of the biosphere in maintaining the livability of Earth systems: "If you lost the animals and plants, you might lose the speed, but you would never qualitatively lose the cycle. If you lost the microorganisms, you'd lose everything; you'd unhinge the articulations of the biosphere." Yet microbiology is still largely ignored by the other disciplines.[82]

A fascinating scientific frontier right now is the manipulation of genetic characteristics. Hubris dies hard. It is easy to believe that we can manufacture nature to our specifications. We forget that we are not creating those characteristics; we are just moving them from one organism to another. As we wipe out the species that can provide them, we lose genetic characteristics that we could have exploited to outwit pathogens and agricultural pests. We need biodiversity in a very practical sense. We need to provide the conditions for other species to survive, even if we have not yet realized their importance to us. The importance of preserving gene pools was not even an imaginable argument a short century ago. What else lies out there that we cannot yet imagine?

That may be the most important long-term pragmatic argument for preserving other species, but there is a tougher question: Should we be impoverishing the Earth of other species, even if they do us no good? The answer to that question lies beyond proof, but I would answer "No, unless they constitute a clear present threat such as an epidemic." It is not just that we may guess wrong. A diverse Earth is a wonderful system.

We destroy most species by destroying their habitats, rather than by simply killing them. If we can preserve the room for them to live without interference, we will help to preserve other species. (Not all of them. Species have disappeared almost as fast as they have appeared, long before humans arrived, but not at the present rate.) The UNESCO (United Nations Educational, Scientific, and Cultural Organization) Man and the Biosphere Program and its Biosphere Preserves were created a generation ago to provide that space. They have been in trouble, partly because they are costly but mostly because it is politically impossible to enforce the sequestration of such areas when expanding human populations need the land or the resources. Recently, farmers in Uganda murdered four great apes of the 300 remaining in the Bwindi Forest National Park—half the world's remaining population. (This happened despite a project to protect the park against poaching by enlisting locals in its management.) The project manager said that the area "is full of people, . . . and you are living in a delusion" if you think you can keep them out.[83] (I would add that it is a lot fuller now than it was a few years ago.) We are on the way to wiping out some near relatives.

The lesson, of course, is that one cannot hope to save the diversity of species in competition with a human population expanding to occupy the terrestrial globe (p. 20). We will continue to wipe out other species and perhaps inadvertently threaten ourselves unless we deliberately set out to preserve the diversity of life. This change of direction would require more than larger preserves; it would demand a habit of thought that would warn us when we might be endangering other species, including those we cannot see, and a recognition that the entire Earth system is, in effect, a preserve. Designated preserves are not enough when the alteration of habitats is worldwide, as in the case of climate change or UV radiation. There is no place, as I said, for plankton to hide. Or, eventually, for us.

■ ■ ■

This calls, not for palliatives like "slowing population growth," but for a much smaller human population—perhaps the 2 or 3 billion that may also be a condition for avoiding climate warming (p. 75). There were "only" 2 billion of us in 1930 and 3 billion in 1960. Perhaps we passed the right level somewhere in that period. We cannot prove an optimum at this stage in human thinking, but those numbers are not simply fanciful.

MUTATIONS: ENCOURAGING OUR ENEMIES

There is one sort of diversity we don't like: pathogenic mutations. In chapter 2 I described the forces that drive us to rely more and more on the highest yielding cultivars and the ways in which that very practice causes pests to mutate and become resistant to pesticides. The harder we fight them with pesticides, the more we encourage dangerous mutations. On the other hand, we need the monocultures to feed a growing population so long as we stay in the demographic squirrel cage.

Human diseases go through a similar response cycle to the drugs we use against them. The process has been accelerated by the explosion of diseases caused by the poverty, crowding and hunger resulting from population growth and uncontrolled urbanization of the third world. We are engaged in an endless battle. It would be easier if we did not unconsciously encourage our microbial opponents by perpetuating the conditions in which they mutate and thrive.

DISCONTINUITIES AND SYNERGISMS

We tend to project the future as a linear extrapolation from the present except when it is too painful. Then we fudge it. (Economic and demographic projections are regularly fudged if the model leads to "unrealistic" conclusions.)

The standard formula much used by population writers is Environmental Impact = Population x Consumption x Technology. Simply expressed: $I=PCT$. That is a linear formula. It assumes that all three factors move in straight lines.

There are potential nonlinearities that could make things much worse. They are worth noting, gloomy as they are. In a recent article, Norman Myers catalogued several of the more serious candidates. He divided them into *discontinuities*, which are major qualitative changes when a system is stressed, and *synergisms*, which cause one environmental process to reinforce another. The possibilities include the following:

1. a discontinuity—a shift in the Gulf Stream caused by climate change—leading to the collapse of European agriculture, perhaps intensified by the synergistic effects of soil erosion, low–level ozone, acid precipitation, increasing ultraviolet radiation, and/or the multiplication of pesticide-resistant agricultural pests. (This is not as fanciful as it may sound; the IPCC warned of possible "surprises" in North Atlantic currents. [See page 63.])

2. the synergistic effect of a growing population gathering firewood from diminishing forests once the harvest passes the replacement level, or the similar (and now familiar) accelerating decline of ocean fish stocks once the sustainable catch is exceeded.

3. the synergism of CO_2-induced global warming causing increased ultraviolet radiation (via stratospheric ice clouds and ozone depletion), thereby reducing the oceanic phytoplankton that presently absorbs CO_2 (see chapter 3), leading to more global warming, and so on.[84]

These are warnings rather than predictions. The Earth has been a remarkably stable system, and we tend to take it too much for granted. Nonlinearities occur all the time (see the discussion of urbanization in chapter 1). Myers is trying to tell us that there are some potential ones in the wings that could have immense consequences. That possibility alone is an argument against pushing our disturbance of the biosphere too far.

The more issues we look at, the clearer the margins of growth become. A wise man once remarked to me that it is remarkable how stable systems can be, until they go beyond a certain point; then they can be remarkably unstable.

■　■　■

This is becoming quite a litany: add ultraviolet radiation, waste production, unemployment and now biological impoverishment to the list on page 76. We damage our little spaceship even as we demand more of it. And it is quite possible that a combination of such insults will create a sharp break or an accelerated decline of the support systems.

It will take a basic rethinking of the relationship of humans to other species and indeed to the Earth itself before we are ready to act deliberately to preserve our vessel. Walter Lippman once remarked that diplomacy is the art of matching one's commitments to one's power. Substitute the word "resources" for "power," and the statement becomes true of conservation—and much more important. We must look at the system as a whole, recognize its limits, see the circular flow of resources and "wastes" and realize that at some point—apparently in the past—human demands simply get too large. If I may approach the edges of the mystical: we are ourselves more whole and balanced if we understand our role as part of the web of the biosphere rather than imagining ourselves its conquerors.

PART II

Interlocked Consequences: A Darkening Century

THE NEXT HALF-CENTURY seems to be shaping up as the age of mismatch between hopes and the attainable.

What seemed two generations ago to be the best and most promising of times has shifted beneath our feet by almost imperceptible stages into a world of rising poverty, pollution, social unrest and deep fears and uncertainties even among the privileged of Earth. What has happened? How do the driving forces described in Part I play out in countries at different stages of development, and how do those disparate futures play against each other? What can be done about them?

CHAPTER 7

Diverging Futures

THE JUGGERNAUT, like Proteus or Vishnu, seems to have different shapes. In the poorest countries, it is sheer population growth, unmitigated by rising consumption.

The emerging third world countries are generally thought to be on a roll, but they face the danger-laden combination of sharply rising population and consumption. The most successful ones will be those that have stopped population growth and brought most of their workers into the modern era; others like India and China will be riven by a division between those in the modern sector and the ones left behind. Both kinds will by their very success multiply the environmental damage that industrial countries are already generating. Their success will be based for the foreseeable future on cheap labor, and they are already contributing to wage, welfare and unemployment problems in the richer countries.

The rocket of rising crop yields will probably slow down, just as the newly industrialized countries—almost none of which can meet their own food needs—become able to compete for food on the world market. Total world food trade may not be enough to meet their needs.

The old industrial countries face the problem of sustaining high consumption and employment in densely populated societies, in the face of competition from the emerging countries. Their problems would be relatively manageable if the rich could somehow isolate themselves from the rest of the world, which they can't.

To simplify an enormous complexity, let us look first at Africa as a study of the poorest countries that show no signs of escape from poverty, and the conclusions that one must draw from it. Then let us look at China—itself the home of one-fifth of the human race—as an example of the poor countries that are industrializing, but by no means uniformly, and then at "other Chinas" following similar paths. Finally, we will consider

Europe and the "old industrial countries," with problems arising, not from indigenous population growth, but from their high consumption levels and from the consequences of population pressures elsewhere.

AFRICA: THE DESPERATE CONTINENT

It takes strong nerves to contemplate the next half–century. Population momentum very nearly guarantees that the poorest countries will become more desperate. They are in danger of becoming marginal to world trade and world concerns, but in an interlocked world we will not escape their afflictions.

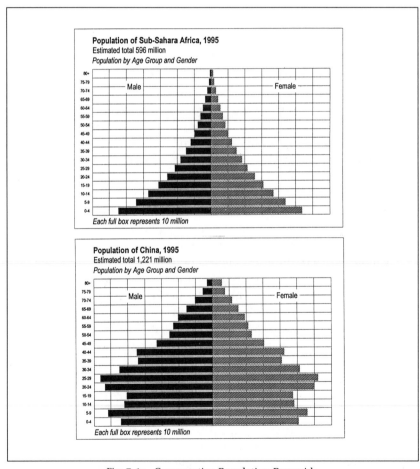

Fig. 7–1a. Comparative Population Pyramids
(*UN 1994*)

Africa is the first region reaching its population limits, along with Haiti and parts of south Asia. News stories from Africa dramatize the appalling reality of that condition. In the case of most sub–Saharan Africa, the constraints are not absolute but rather are the product of extremely fast population growth pressing so tightly against present resources that there is little room for maneuver, compounded by political turmoil and racial tension that make it very difficult to plan for any future beyond a few weeks ahead. Those countries are trapped. Arable land is badly used and underproductive, but they have neither the funds nor the slack to take the risk of modernizing their agriculture; the penalty of failure would be immediate starvation. Their governments do not have the depth of

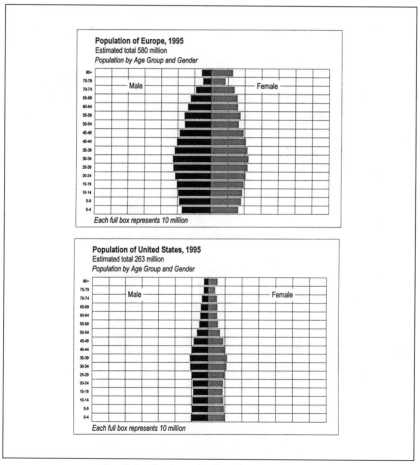

Fig. 7–1b. Comparative Population Pyramids
(UN 1994)

administrative structure either to stop or adjust to population growth that is out of control, and population momentum is driving them into an impossible corner.

The International Labor Organization speaks of the "collapse" of the modern sector in Africa. Unemployment is rising and real wages are falling—by 50% in Kenya since 1975 and 80% in Tanzania. Foreign investors, naturally, stay away. Only 0.6% of private foreign investment has gone to all 47 of the poorest countries.[85]

The present age distribution of Africa's population is compared with China, Europe and the United States in figures 7–1a and 7–1b to dramatize the utterly different population issues the different regions face.

To demonstrate the consequences of those different pyramid shapes, let me project the total population of each of those areas to 2050. I have used *constant fertility* projections, because they are the easiest to understand; they simply show what will happen if fertility does not change.

For China, Europe and the United States, the UN constant fertility projection is not much different from the *medium* projection, which is usually taken as the UN's estimate of the most likely one. The *low* projection would lead to stabilization of the Chinese and U.S. populations at about 1995

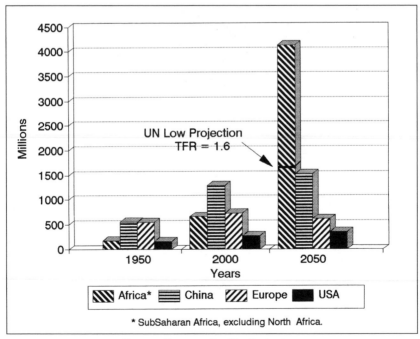

Fig. 7–2. Constant Fertility Projections
(*UN 1994*)

levels, with Europe's population declining faster. For sub-Saharan Africa, however, the difference between the constant and low fertility projections is 2.6 billion people. That is the measure of what a fertility decline to 1.6 could achieve. Unfortunately, rising death rates are more likely to stabilize that population. Even the "low" figure represents nearly a trebling, and I cannot believe the continent can support it.

What do we learn from those graphs? First, population momentum: those huge cohorts of young African girls will be having children in years to come. Even if they should average only one-third as many children as their mothers are having, their sheer numbers would drive population upward for generations. There is no relief in sight.

Second, the burden of raising so many children diverts resources that might have been used for investment in a better life. The "dependency ratio" is the number of young and old per 100 "working age" people 15–64 years old. One hears regularly about the high dependency ratio of aging societies like Europe or the United States. In Africa, because there are so many children, it is much worse: 89 per 100 working–age Africans compared with 49 for Europe.

Third, there is a huge problem of helping those children find work as they come of age, and the number will double in the next 20 years—if enough of them survive.

Finally, that huge and growing population, seeking something to eat and something to do, will put enormous pressure on the land and the remaining forests in Africa. There is now just over a quarter hectare of arable land per person in sub-Saharan Africa; it is declining rapidly and by 2030 will be about 0.1 hectare, a level comparable to those in present-day Bangladesh and China. African land is fragile and highly susceptible to droughts and desertification; two-thirds of it is classified "arid," contrary to the popular image of vast jungles.

There is, in a sense, a bit of room—if it can be exploited. Agricultural yields in sub-Saharan Africa rose about 60% between 1950 and 1975 and have been static since then, at about one-fourth European or American yields. They could be brought up if somehow modern inputs and methods could be applied, but one should not hope for too much because the land has been so badly misused.

I suspect that the statistics on African food supplies are little more than guesses, and they seem to be internally contradictory. Perhaps the simplest indicator is a World Resources Institute (WRI) calculation taken from FAO data: food production *per capita* in Africa declined 20% between 1970 and 1992.[86]

The USDA has estimated emergency world food needs, mostly in Africa, at 15 million tons at the present, and 27 million tons in 2005, even without a production crisis.[87] Food aid, mostly from the United States and Europe, has declined from about 11 million tons annually to a predicted 7.6 million tons in 1995/96 despite pleas from the FAO, and it is unlikely to rise for reasons that will become clear in subsequent pages. Africa for several years has suffered combinations of drought and hundreds of thousands of refugees. Starvation is probably a present reality, not just a threat, but we don't know the numbers.

The final irony is compounded by agricultural realities and demographic laws.

- If those countries tried to modernize agriculture in the conventional way and succeeded, they would pollute their water supplies and drive themselves into the nitrogen trap described in chapter 2, unless those babies stop coming so fast.

- Given that astonishing population pattern in figure 7-1a, Africa would need a family planning program more drastic than China's to break the cycle of desperation. In sub-Saharan Africa, only South Africa seems to have the organizational apparatus to dream of such a program, and even there it would be difficult in an atmosphere tense with racial rivalries.

- There is a Catch 22: even if Africa could achieve the low projection and population programs succeeded fast enough to arrest that awful momentum quickly, Africa would then have a huge population of old people three generations down the road, with very few children coming along to support them.

In 2050 (Figure 7-3), the working-age population would be very large (this pyramid assumes there is no AIDS or other epidemic), but by 2075 those people would be old, and the young cohorts coming of working age would be progressively smaller.

The fatal reality of demography is that one cannot just walk away from the mistakes of the past. Africa's population momentum will haunt it for the next century. There isn't time for a gradual solution, and a swift solution would generate massive problems of its own. The best that can be said is that the low projection future would be better than the alternatives, even though Africa's monstrous "baby boom" would face a bleak old age.

Have I made my point? Africa is already well into a catastrophe, and I cannot even visualize a way out.

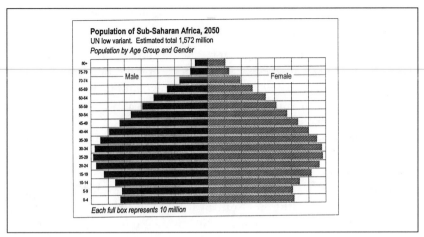

Fig. 7–3. Africa's Best Hope
(*UN 1994*)

THE FADING DREAM

Africa epitomizes the problem of the poorest. Those countries are living through a tragedy. The sub–Saharan African population was about 170 million in 1950, less than one-third its present size. With the new technologies available, such a population could reasonably have aspired to a world in which all could live at a decent level. Instead, mortality was brought down through foreign aid and the introduction of modern medicine and public health concepts, long before anybody addressed fertility, thus generating the population explosion that now makes a mockery of those hopes.

Professor E. O. Wilson has written a bitter statement about Bangladesh that could be an epitaph for most poor nations:

> The raging monster on the land is population growth. In its presence, sustainability is but a fragile theoretical construct. To say, as many do, that the difficulties of nations are not due to people but to poor ideology or land–use management is sophistic. If Bangladesh had 10 million inhabitants rather than 115 million, its impoverished people could live on prosperous farms away from the dangerous floodplains . . . It is also sophistic to point to the Netherlands and Japan, as many commentators incredibly still do, as models of densely populated but prosperous societies. Both are highly specialized industrial nations dependent upon massive imports of natural resources from the rest of the world. If all nations held the same number of people per square kilometer, they would converge in

quality of life to Bangladesh rather than to the Netherlands and Japan . . . [88]

Foreign aid cannot fill the gap it has helped create. Indeed, as aid suppliers are slowly learning, even "emergency" food aid becomes a danger when it becomes a crutch because it keeps an expanding population alive to face the next bad season under worse conditions. It is very hard to withhold food at such times or to demand a quid pro quo such as specific reforms. International organizations, out of habit and compassion, continue to urge more emergency food supplies. However, as world agricultural surpluses dry up, food aid is the first thing to go.

At early stages of human development, before the Earth was so fully occupied, societies could grow out of their problems or flee them—or sometimes simply move onto others' turf, as the Kikuyu in East Africa did for generations. That time has gone. Chaos would follow expansionist policies in Africa. Although most African borders are artificial constructs—a legacy of European colonial struggles—African nations recognize the danger of tinkering with them and have avoided making claims on each other's territory. There are no open frontiers.

There is a name for uncontrolled growth in living organisms: cancer.
Africa is living through its social analog.

Africa is in the classical demographic cycle: new technology leads to enhanced possibilities, which leads to population growth, which leads back to poverty. It is demonstrating what happens when growth passes the ability of the ecological and social systems to absorb it: destruction of resources, poverty, unemployment, rural desperation, migration to the cities, chaos in the overloaded cities, and a reversion to tribal warfare such as we are witnessing in much of the continent.

There is nobody to blame. The rich countries had gone through the industrial revolution, when everything seemed possible, and the poor countries thought they could follow suit. Fossil fuel and the dreams of atomic power and "energy too cheap to meter" seemed to offer an inexhaustible energy fix. Growth seemed limitless. In an era of optimism, one tends to plunge ahead without considering the consequences.

The UN classifies much of sub-Saharan Africa plus several countries in South Asia and Latin America as "least developed." Those poorest countries constitute just 10% of world population now, but 26% of world population growth to 2050 will take place in those countries, which will treble in numbers (UN 1994 median projection)—if they can get there.

I don't believe they will get there. Their population growth is approaching the point at which it will be stopped or reversed by the Four Horsemen—war, hunger, pestilence and a rising death rate. Most of us once thought humanity had escaped those miseries. We do not know exactly where that point is, but it will be preceded—it is already being preceded in some countries—by rising misery, particularly among those most fertile.

AIDS is making deep inroads in Africa. Unlike the epidemic in the industrial world, the disease in Africa strikes women about as often as it strikes men. In four countries, more than 5% of the total population is afflicted.[89] At that rate, the United States would have 13 million AIDS patients. AIDS is beginning to have an impact on population growth. The U.S. Bureau of the Census has taken a first cut at calculating the impact of AIDS on population growth in the most afflicted countries:

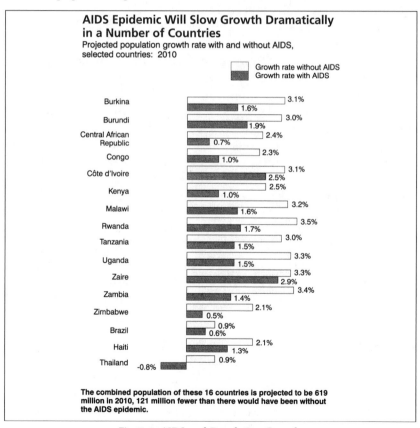

Fig. 7–4. AIDS and Population Growth
(*Census Bureau 1994*)

It has not yet reached a level that would stop or reverse African population growth (although it may in Thailand). The future is unknowable. AIDS has shown a remarkable tendency to mutate and escape efforts to control it, and it is an unpredictable force for Africa and the world. It is a tough way to solve the population problem. Who takes care of all those children (figure 7–1a)? This is a grim picture indeed.

The official estimate of HIV sufferers worldwide is about 22 million, which is said to be a doubling since 1991, and projections for HIV incidence by 2000 call for another doubling or perhaps much more. Active AIDS cases at present are estimated at anything from 1.4 to 5 million, and people are dying at the rate of about 1.3 million annually.[90]

If not AIDS, the Ebola virus? There is a set of afflictions already taking shape: AIDs in its permutations, existing diseases of crowding and poverty, rising losses to agricultural pests, and ominous new mutant pathogens afflicting agriculture and humans.

Pestilence has benefits, in a grim way. In Italy, the Black Plague followed a period of recurrent hunger and lesser plagues, a result perhaps of population growth and/or climate change. The population decrease took the pressure off arable land and led to better and more diversified diets and to a rise in per capita incomes. There is evidence that the improvements contributed to the Renaissance.[91] A sustaining thought, but does Africa have to go through a plague?

Economically, those poorest countries are simply fading out of the world scene, and Western compassion fatigue slowly erodes our interest in them. This does not mean that somehow the world is rid of them. We are linked to them by epidemics and migration (chapter 8).

THE "EMERGING" NATIONS: CHINA AND THE WORLD

China is not just an example of an industrializing nation. Alone, it contains one–fifth of the human race, and every nation has a stake in its future. The Chinese are cursed if they try to master their problems, damned if they don't, in trouble if they succeed and in worse trouble if they fail. Let me explain.

The Race Against Hunger

The Chinese have a massive population program underway, epitomized in the slogan "the one–child family." Unlike most poor countries, they are financing most of it themselves. The program involves very intense social and economic pressures on women not to become pregnant after the first child and to abort if they do. There are various exceptions involving minorities, twins, and peasant families with labor shortages. The program—like most campaigns in China—is not so totally effective as propaganda would suggest. UN figures show total fertility in China right now as 1.95 children. The head of China's family planning program has said that the official data probably understate fertility.[92] The program has been bitterly criticized abroad, particularly in the United States, for coercion. It has in fact had its excesses, as have many of the campaigns launched in China.[93] Yet China cannot afford to give it up.

Let's look at their problem.

China's population is 1.22 *billion*, growing just over 1% per year. The rate of growth is similar to the United States', but in China that works out to more than 12 million people every year.

Traditional agriculture in China could achieve food yields sufficient to support as many as seven persons per hectare of arable land on a simple vegetable diet, but in only a few favored areas, under ideal conditions and with massive human labor, legume rotations and the recycling of manure and nightsoil.

The population per hectare is now 13 persons and rising.[94] And those people have gotten used to having a little meat, rather than living at or below the margin, as they used to, on a grain diet supplemented by a few vegetables and some vegetable oil. China escaped from subsistence agriculture by building fertilizer plants starting in the mid–1950s. It has almost reached the end of that road. China uses nearly three times as much fertilizer per hectare as the United States, but grain yields are only 87% as high, probably because they must farm poorer land in drier areas and have less experience with modern high–yield hybrids and crop varieties. In some areas of heaviest application, the response to further fertilization in China is zero.

Those are the limits that every nation has encountered when it moved toward massive use of artificial fertilizer. China's situation is particularly tight because it has so little arable land. It cannot go back to organic agriculture with its much lower average yields. Vaclav Smil (chapter 2) points out that synthetic fertilizer furnishes roughly half the nitrogen in Chinese food production. Even if China reverted to subsistence consumption levels, commercial fertilizer would be needed for 40% of the essential nitrogen. Despite its harmful environmental effects, it is essential. Without it, China would face mass starvation. Since China has long farmed substantially all its arable land and is using the available sources of organic nitrogen, including nightsoil, any future increase in Chinese population will depend largely upon synthetic nitrogen, at a stage when the added nitrogen produces very little food.

To add to its woes, China is losing topsoil to erosion and farmland to other uses. A Chinese official told an environmental meeting that one-third of China's arable acreage is seriously eroded and it is losing farmland to urban and industrial expansion and the housing needs of a growing population. Official figures show a decline of 6% in arable land from 1970–1991 and a loss in 1992 to urbanization alone at a rate equal to nearly 5% per decade.[95] Those losses are particularly significant, since the land closest to cities gets the most nightsoil and is highly productive.

Irrigation is pushing its limits. Consider one example: some 622 kilometers of the Yellow River went dry for nearly four months in the summer of 1995. It was a dry year, but the phenomenon was blamed in part on a nine-fold increase in irrigation from the river since 1950.[96] Around Beijing, irrigation is being reduced to meet urban water needs.

Every growing population is on an agricultural treadmill, but China is the ultimate example. The Chinese are in a race against time, and they know they must solve their own problems. No supplier on Earth could feed them if they fail. Their birth control program must succeed or they will face a descent into starvation.

Those who criticize Chinese family planning policies should perhaps consider their problem. As the Chinese premier bluntly told ex-President Bush: "If China's population goes on increasing uncontrolled . . . China will land in a backward state and even head for self-destruction."[97]

China as Importer
China is likely to be a major entry into the world market, particularly in two areas.

Grain. China is entering the grain market on a massive scale just at the time when stagnating yields and competition from other emerging countries are making that market very tight.

In the bad old days, China could not afford imported grain. If their crops failed, people starved. Then, as commercial fertilizers became available and food output rose, China briefly became a food exporter. That has changed again, and China now runs a net grain import balance around 15 million tons or more, second only to Japan. As a result of its relaxation of controls and its economic boom, China can afford to buy grain. It needs to for domestic reasons. The price of grain rose more than 50% in 1994.[98] That is the kind of thing that leads to food riots.

There has been something of an international intellectual skirmish as to how much grain China will need a generation hence and where it will get it. The Worldwatch Institute projects a requirement for 479 million more tons by 2030—of which 207 million tons would have to be imported—even if improvements in the Chinese diet should stop now. If the diet continues to improve to something like the level in Taiwan, Worldwatch projects the need at 641 million tons, 369 million tons of which would have to be imported.[99] That is nearly twice the present level of total world grain exports.

China's Agriculture Minister responded with an ambitious goal: avoid imports by raising grain production to 500 million tons in 2000 and 625–675 million tons in 2020—about 45% above current production. As the cynical would say: lots of luck.

Other specialists, Chinese and foreign, offer a wide range of projections of production and import requirements early in the next century ranging from zero to 136 million tons per year.[100] Nobody really knows, but nobody is claiming China will again be a net exporter.

One does not need to take the astonishing Worldwatch Institute projections literally in order to agree with the thrust. Worldwatch assumed a decline of 20% in Chinese grain production as a result mostly of declining acreage, less irrigation and stagnating yields. That may be pessimistic, but the Minister of Agriculture is extremely optimistic.

Chinese imports will depend on how much they can afford, i.e., how much foreign exchange they can muster and the price of grain. I read the Worldwatch message as saying, not that they will buy that much, but that they will not be able to, because the world's importers will be competing bitterly for a limited supply (chapter 8).

If they cannot raise the grain or import it, the Chinese government will have to tell the Chinese to tighten their belts. A lot of peasants may be reluctant to part with their grain and insistent on sharing in the benefits of a rising market for what they do sell. A very tough spot for government and one that leads me to repeat the refrain: China's problems must not be worsened by rising population–driven demand.

If China had already succeeded in its hope of reversing population growth, the specter raised by the Worldwatch projections would have disappeared. Population growth is the problem on both sides of the calculation. It raises demand while it competes for farmland and requires more and more water from limited resources for irrigation, more housing, larger cities and expanding industries.

Petroleum. Something of the same calculus applies to the world petroleum market. The Chinese cannot meet their present needs. The Ninth Five Year Plan (1996–2000) identifies grain and petroleum shortages as two of the most serious problems China faces. Petroleum production has stagnated for six years, despite the economic boom. South China in particular is very short of energy. China has become a major petroleum market in the world. China's State Planning Commission admits imports will rise but not "substantially."[101] I suspect it is optimistic. China's purchases will rise aggressively if it can afford them. China cannot turn to coal; it already supplies three-quarters of China's energy and ties up nearly half of its inland transport capacity. China cannot presently move or burn much more. A significant role for nuclear energy is decades or generations away. One expert predicts China's petroleum imports will rise four-fold to 1.4 million barrels per day by 2005.[102]

Imports at that scale would absorb over 3% of world exports (1992) and would compete with rising demand from most of the world. They would hasten the energy transition for the whole world. Over time, if the Chinese cannot buy the petroleum, they will rely on coal and nuclear energy, with profound environmental consequences (chapter 4). Damned if they do, damned if they don't.

Industry and the Peasants

There are at least two Chinas. Not just the familiar China versus Taiwan division, but the far more fundamental fission line between the prospering southern coastal cities and the peasantry, particularly in China's vast interior. Only a small part of China is participating in the present economic boom. The peasants know that some Chinese are getting rich, while the peasants have financed the cities through grain levies and taxes. There have been riots and attacks on local government offices. It is not comfortable out there in the countryside.

That division has far-reaching implications. A restless and growing peasantry is in danger of losing the gains of the past 30 years, and the loss of expectations leads to a dangerous mood indeed. They are running out of farmland and crowding into overloaded cities, encountering a small business plutocracy, some prospering entrepreneurs and a small group of skilled workers who are "in the system."

The Ninth Five Year Plan addressed the problem with considerable candor. Rural and urban unemployment and the "polarization of rich and poor" were identified as critical problems. One drafter, in presenting the draft plan, said, "Unemployment was the deciding factor in setting the growth rate . . . The only way to keep unrest from boiling over . . . is to create a certain number of jobs."[103]

The official figures on unemployment are indeed hair-raising. Outside the cities there are reported to be 450 million rural workers, of whom 120 million are employed in rural industrial enterprises. Agriculture can use 200 million at most, leaving 100 to 200 million already surplus (by various officials' estimates). Estimates of unemployment originating in the countryside alone range from 180 to 300 million by 2000.[104] Worse still, the government hopes to modernize farming along Western lines and proposes over time to shift 80% of farm labor into other work, and their job prospects will be dim. The labor force grows about 10 million every year, and the cities will also generate substantial unemployment as the government tries to reduce the padded rolls of state enterprises.

This nightmare is Ricardo's Iron Law of Wages, driving wages toward minimal subsistence levels. Quite reasonably, the poor want out. Witness the desperate efforts by Chinese to get to the United States even in the midst of a touted economic boom—and most of those boat people come from Fukian Province, which is supposed to be prospering. There is even a governmental "overseas employment agency" in one depressed northeastern city to help Chinese to emigrate.[105]

Because China's family planning program has succeeded pretty well, there is hope down the road. The number of young people entering their working years is already below the 1985 figure. It will fluctuate downward if China can hold its fertility down,[106] thus ameliorating the unemployment problem. Those smaller cohorts will be entering the labor market in 15 to 20 years (see figure 7-1a), and future cohorts will be still smaller as that present 20–34 year bulge moves out of their childbearing years—if China can hold fertility below replacement level. That sudden tuck in the bottom of the Chinese pyramid, compared with the flaring African pyramid, means fewer children to support now, an eventual reduction in the huge number of unemployed, fewer mothers having babies in the next generation and the hope of stopping runaway population growth. The comparison with Africa's plight is a dramatic argument for population programs.

Industry and the Environment

Like most other people, the Chinese do not simply want to stay alive. They want to succeed, and the economic liberalization that Deng Xiaoping launched has created a class of people for whom that dream has become possible. China epitomizes some uncomfortable realities. The poor want to be rich; poor nations want to live like us.

China has one of the world's fastest growing economies. GNP grew from 1991 to 1994 at an average annual rate of 11.7%. They plan a "more balanced" growth of 9.3% through 2000 and 8% from 2001 to 2010. That would mean a seven-fold increase in 25 years. They hope to reach the per capita GNP level of a "middle income developed country" (i.e., the OECD median) by midcentury.[107] Allowing for projected Chinese population growth (UN 1994 median projection), that target is more than 70 times as large as their 1990 GNP—even assuming OECD GNP stands still. (That astonishing target is not unique. Remember that the IPCC assumption for the whole third world is a 70-fold increase by 2100. [See p. 69.] The Chinese are just in more of a hurry.)

They are not likely to get there, but in the process of trying they will generate some profound disturbances to themselves and the world. Their problem is that, having abandoned the Communist model (in fact if not in word), they have adopted the capitalistic model—conspicuous consumption and all—just as it begins to run against the realities of a small Earth. China produced 10,000 automobiles in 1985 and 1.4 million in 1994. It hopes to treble that rate by 2010. By then, they expect to have 40–50 million private automobiles.[108] They could hardly have selected a worse way to go. Automobiles eat up land for roads and parking and the urban layouts they promote, and much of that land will come out of China's dwindling farmland. Their pollution may harm farm production; it will contribute carbon to global warming, which will be a disaster for low-lying coastal areas. It will imperil food production if it leads to less rainfall in a hinterland already suffering from lack of water (see p. 66). Their demand for gasoline will consume foreign exchange that the Chinese will need for food imports.

In retrospect, it may seem a tragedy that the Chinese leaders, given the chance, did not set out to define a more benign course of modernization when they abandoned the Maoist rigidities. One could envisage a China built upon freer enterprise and the price mechanism but organized around the countryside and smaller cities, emphasizing labor-intensive occupations and avoiding the energy-intensive Western industrial models. Apparently, the leaders did not see the vision or did not feel they had a choice. Chinese want the things they see on television, which now reaches most villages. If their government fails in its ambitious plans, it will face a restive populace. If it succeeds, China may well choke itself—and us—on pollution.

As with other nations that have been poor, China's leadership does not yet appreciate the penalties that growth will bring. Already, 86% of rivers in urban areas are regarded as heavily polluted. Shanghai alone dumps 5 million cubic meters of raw sewage into its rivers daily.[109] Unwashed coal provides most of China's energy. The air is foul. Entrepreneurs are even setting up "oxygen kiosks" in some cities, where people can stop for a whiff of pure oxygen. Despite some help from the World Bank

and the Asian Development Bank, Chinese environmental leaders themselves say that China cannot afford to do all that needs to be done.

China will not suffer alone. Already, acid precipitation from China is
affecting western Japan. China has every intention of using its coal
resources, and coal is the dirtiest fuel. There are technologies that could
help to lessen the damage, but they are likely to be far more expensive
than conventional coal power plants (chapter 4). The industrial world has
an interest in helping China avoid inflicting irreparable damage on the
atmosphere. It could make those technologies available on concessional
terms if China could be persuaded to use them.

There are other limits. Some gain in controlling pollution would be
achieved by washing the coal or using scrubbers—but most of China's coal
is in the dry northwest, which already has water problems. This is one of
those vicious circles: water shortages add to China's problems in cleaning
the coal, but pollution from dirty coal may well make the water problem
worse.

In China most coal is used in small home stoves that are particularly
dirty. They can be improved, perhaps by adding limestone to the coal bri-
quettes, but they will remain a dirty source of energy.[110]

China as Exporter

The immense and underutilized pool of cheap and trainable labor will
keep Chinese wages low. If Chinese labor is mobilized, whether by Chinese entrepreneurs or multinational corporations, and if they can get their
goods into foreign markets, the Chinese will drive wages down in labor-
intensive industries wherever they have unrestricted entry. If they are
closed out, the Chinese government will be sitting on an immense pow-
der keg of people with nothing to do.

The United States has been providing that market, which grew from
$1 billion in 1980 to $39 billion in 1994. The United States' China trade
deficit is second only to that with Japan. The United States will probably
begin to recover the deficit as China's grain needs rise, but the depressed
state of American employment will probably generate increasing pressure
to limit access to the American market, as we shall see later.

The Chinese need us much more than we need them, as a market and
a source of food. As the proud heirs of "The Central Kingdom," they
undoubtedly resent it. This makes our politicians' moralizing to them
about personal freedoms and their population policy particularly galling.
They pretty much have to take it, but our politicians are building a
tremendous tension of suppressed resentment. It is remarkable how
insensitive nations can be to each other's needs.

However that may be, the central trade issue is that China needs trade opportunities that the rest of the world, in attempting to save its own economies, may not be able to provide. Again, as with so many other things, population plays a central role. The Chinese are trying to work their way out of the box created by a growing labor force, and figure 7-1a suggests they have a chance. Let us not make it more difficult.

The Nervous Giant

China may seem huge and invulnerable from the outside, but it is a nation trying to survive on very treacherous ground. The question is regularly asked: "what will China do with Hong Kong after it takes over in 1997." Perhaps a more important question is what Hong Kong will do to China. The Chinese leaders don't trust democracy. For a century they have tried to accommodate their ways of managing a huge nation to the needs of modernism. Confucianism was their traditional political framework, and it didn't work. The experience of the Soviet Union and Eastern Europe has made clear to them that statism doesn't work. Communism as a belief and unifying force has eroded in China just as it did in the rest of the Communist world.

China traditionally has broken apart when strong central government failed. Its leaders are intensely aware of that tradition. With the gradual passage of Deng Xiaoping from the scene, the issue of succession is at the forefront of Chinese politics. Communism does not provide a reliable way to manage successions. This is a critical time for China, and today's policies are not perpetual. One issue may become crucial: the one–child policy was not introduced because Chinese like it. The leaders knew they had to stop population growth, even at considerable political risk, to avoid future catastrophe. In the present power struggle, will they be able to stick to that policy, or will one or another leader, courting popularity, propose to revoke it? Or will it simply become unenforceable if central authority weakens? Population policy has shifted before in the political seesaw since the idea of family planning was first introduced in the mid–1950s. There are ominous reports of rising resistance to it in the countryside.

We should perhaps consider that context when we evaluate the violent Chinese reaction when Hong Kong Governor Patten began to introduce direct elections into Hong Kong not long before its 1997 reversion to China. The Chinese were not just throwing their weight around. Patten was introducing a democratic precedent that the incorporation of Hong Kong into China will inject into the body politic, and the Chinese leaders are probably terrified at the injection.

China faces an uncertain future. Some people perhaps would like to see it fail. I am not so sure. Aside from the human aspect, which is important, one can hardly be comfortable with the prospect of chaos and hunger in that much of the human race, in a nation possessing modern armaments including atomic weaponry.

On the other hand, if China succeeds, it will tighten an already tight world grain supply, hasten the end of the petroleum era and multiply the insults the human race is already inflicting on the environment. Environmentalism has been low among China's priorities, and it will take some tolerance and understanding to bring them to agree that they share our interest in protecting the environment.

Migration

Fail or succeed, China will probably generate migration. If they fail, there will again be hunger in China, and that raises the prospect of hungry Chinese pushing into any area where they can find land and food. The Chinese have not traditionally been thought of as a nation of migrants, but that is a simplification. They have been pushing and drifting southward for over two thousand years. With a very few exceptions, they have not mounted invasions when they encountered a sufficiently organized non-Chinese society, such as the Thai or Vietnamese, but the informal flow of migrants into Southeast Asia continued. There are something like 30 million "overseas Chinese" in that region now. One may assume that chaos in China would aggravate that flow. There is a similar flow across the border into Siberia, even though the Chinese and Russian governments have an agreement to control it. The Russian Defense Minister has told the Russian Cabinet that "Chinese citizens are peacefully conquering Russia's Far East." The Cabinet concluded that some of the ethnic Russians coming back to Russia from the erstwhile Soviet empire should be resettled in the border region as a buffer.[111]

On the other hand, economic success will provide the funds to finance boat people to more distant lands. The movement has quite a tradition, starting with the movement of Chinese from China's south coast near Canton to America in the nineteenth century. Fukianese boat people recently apprehended in the United States said they paid $10,000 or more for the trip. One needs relatives with enough money to bankroll that sort of movement, and candidates to undertake it. The combination of economic growth with widespread unemployment is precisely the matrix to support such a flow.

What Is at Stake?

What does that all mean for the rest of us? We may see China becoming several things at once:

- a new and aggressive exporter of low-tech, high-labor content goods, utilizing a vast pool of disciplined and trainable labor, and by their competition depressing world wages,
- a major new bidder in world food and energy markets,
- a desperately polluted coastal zone, adding substantially to the world's atmospheric pollution and contributing to global warming,
- an impoverished, hungry and unemployed peasantry,
- a rising source of migration, and
- a region of instability comprising about one-fifth of humankind.

■ ■ ■

Where does the logic of these trends force us? The only solution that cuts across all those problems is the one the Chinese are already trying: an end to population growth. We have every reason to applaud and support China's desperate efforts to bring its population under control, rather than engage in the constant warfare against it that has marked American and particularly Republican policy for more than a decade. Our criticism reflects domestic politics. We should, amid our fixation with the abortion issue, look at the consequences for China and the United States. "Right-to-lifers" should recognize the costs of their advocacy. There are few if any moral principles as absolute and overriding as their more fervent adherents believe.

We have a stake in China's population program and an interest in helping the Chinese to find the most benign possible ways of making it work. If they fail, and their modernization effort goes with it, they may not add so much to air pollution and climate change, but their government will answer to a billion very unhappy people, and China could become a dangerously unstable player on the world scene.

If only from the standpoint of diminishing the flow of illegal immigrants, the United States has an interest in Chinese success in stopping their population growth.

THOSE "OTHER CHINAS"

To China, add India, Indonesia, Brazil, Mexico, perhaps Pakistan, the Philippines and Thailand, plus the smaller "emerging market economies" like South Korea and Taiwan. Their economies are growing and with it their energy requirements. Their food needs are multiplied by population growth, and they will be low-wage competitors on world markets.

Those countries are by no means replicas of China. India, for example, is just at the edge of modernization. It uses only one-quarter as much chemical fertilizer as China, but it has twice as much arable land per capita. Partly because the green revolution in India was concentrated in a few areas of commercial farming, India intermittently had a small net export balance in cereals in the 1980s. At last reading (1991), however, it was only 91% self-sufficient in cereals, and it is heading for trouble. The total fertility rate is still almost four children per woman, and India's population is expected to rise 42% by 2020, even with some optimistic assumptions about declining fertility. In other words, it has more room for improvement than China, but it is using its slack the wrong way, by absorbing it through population growth.

Those ten countries named above will have three times the population by 2020 that the developed world has today. Development pollutes. Their development will almost certainly precipitate the huge resource and environmental problems that are now surfacing. To give one dramatic example: South Korea (population 45 million) consumes 14% more petroleum than India (population 936 million). What will happen if the third world really industrializes?

The "Brundtland Commission" concluded that, in fairness, we must anticipate a five- or ten-fold growth in world industrial output to accommodate the modernization of the less developed countries.[112] The IPCC, as we have seen, forecasts a seven-fold rise in world GDP by 2050 and 25-fold by 2100. Those estimates epitomize the problem of trying to relieve poverty for a growing world population. It is inconceivable to me that, even with the best efforts at pollution control and conservation strategies, such a growth would be environmentally tolerable, and there is yet no sign that the emerging countries will make their best efforts. Take the greenhouse effect as an example: the developed countries are trying unsuccessfully to hold CO_2 emissions at the 1990 level, the emerging countries won't commit themselves, and any increase will intensify climate warming (chapter 4).

A similar prognosis holds true for acid precipitation, nitrogen production (chapter 2), forest resources, water availability in dry regions and indeed most of the environmental issues touched on in this book.

The newly prosperous countries will be competing for food in a world market that cannot accommodate them (see chapter 8).

Food, I have already pointed out, is connected with population in a particularly inelastic way. Those countries must stop population growth to avoid the looming food crisis.

They will also need more energy to fuel their industry and to meet the expectations of their people, but this brings the world head-on into the fossil fuel issue. The U.S. Energy Information Administration points out that China and India, although they represent only 12% of total world energy demand, "accounted for over 30% of the world's increase in energy use between 1980 and 1993."[113] That sort of growth will push the energy market very hard and hasten the end of the fossil fuel era.

Let us use Asia (excluding Japan and the former USSR) as a surrogate for the emerging nations. Those nations presently use 15% of world energy. The EIA expects their energy needs to double from 1990 to 2010, even with a sharp slowdown in recent growth rates and a dramatic improvement in energy efficiency (p. 47). At that rate, their needs will pass present total world energy consumption in 2045. If they grow as they have been growing since 1970 and don't become more energy-efficient, they will get there in 2027. The present impacts of energy on the environment would be dwarfed by such energy use. Fossil energy resources, particularly oil (chapter 3), cannot accommodate it.

The 2027 date is not just hypothetical. The "efficiency" of energy use is a compound of two things: true efficiencies such as low-energy light bulbs, and the less intensive use of energy as economies mature and their GDP includes more services and a smaller proportion of heavy industry. Although energy specialists are optimistic that the second process is underway in the third world, the evidence so far is mixed. Energy demand has risen faster than GDP since 1980 in some countries, slower in others.[114] Very few of the emerging countries have the completed social infrastructure of the old rich countries; they may take years to become so "energy-efficient."

The EIA has done an interesting little exercise. If the rest of the world had reached the OECD level of per capita energy consumption in 1990—which means both OECD living standards and OECD energy efficiency—total world energy consumption would have been three times the actual level in that year.[115] As a sort of "best and worst case" scenario, population growth alone would run that ratio up to nearly 6:1 by 2050 (UN 1994 medium projection). Not likely, but a useful measure of the upper margin of the problem. If present world energy consumption is already causing the problems I have described, that is how much worse it could get. Some dramatic combination of energy and population policies will be needed to arrest the environmental damage.

The OECD standard of living is the explicit goal of China and the dream of the other emerging countries. The problem is that the dream is unattainable for all but a few, even if there were monumental efforts at efficiency and conservation and a rapid move out of fossil fuels. At some point, not very far away, the emerging nations will recognize that they cannot pursue the growth pattern on which they have embarked, because food shortages, energy scarcity, environmental horrors and eventually the fact of climate change will conspire against them. Or they will be increasingly divided, like China, between a small rich minority and hungry masses. One can hardly predict how those conflicts will play out, but the prospect is for a less and less stable world until growth policies are reconciled with those limits and population is recognized as a key element in the solution.

The World Order Changes

The definition of "industrial" is getting blurred. A Korean industrialist, having established a factory in Antrim, Eire, to serve the European market, remarked that he could pay lower wages there than in South Korea. (He was demanding that the government "do something" about Korean wages.)[116]

Korea is an excellent example of a fundamental dividing line among the emerging nations. Those with fertility below replacement level are on the way into the industrial club—with one qualification. As a rough rule of thumb, the most emergent countries have the lowest human fertility, the most intensive use of fertilizer, the highest agricultural yields—and the greatest dependence on outside sources of food. South Korea has a total fertility rate below China's (or the United States') , which is one reason its wages are rising. It uses more fertilizer per hectare than China and gets higher food yields, but it now has more than 20 persons per hectare of arable land and produces only one-third of the cereals it eats. China is not the only country in the nitrogen trap. Korea, however, has an inestimable advantage: it can reasonably hope to stabilize its population and its food needs around 2040.

Korea is on the verge of joining the "rich" club, and eventually it too will watch jobs move to lower wage countries. The Antrim factory is just a beginning.

THE INDUSTRIAL WORLD

Europe (partly by exporting about 50 million people during the nineteenth century) seemed to escape Malthus' trap and complete its demographic transition. Its problem is not its own population growth, but it shares the Earth with others and cannot escape the changes they are going through. Europe's problems are partly technological (loss of jobs), partly external (low-wage competition) and partly the product of European success (state of the biosphere).

The Joys of Stability

At first glance, Japan and West Europe seem to be in an enviable situation. They are prosperous and at peace. European political unification is not easy, but it is moving forward, and with it come the advantages of free movement, bigger markets and the eventual control of the fratricidal habits that have made Europe a battleground for centuries. Their population growth is coming to a halt.

There are some problems. Europe faces the energy transition relatively poorly endowed with fossil fuels or the potential for solar or wind energy, given its cloudy skies, northern location and low average wind velocities.[117] The potential for biomass is limited, since they are already using the land intensively. However, they have the capital to finance the transition, and with much more efficient use of energy than the United States, they have approached living standards like ours. They are trying out novel sources, such as wave power. (In 1995, a prototype commercial wave-powered generator was installed off Scotland.[118] The first major storm tore the experiment apart. I assume the experimenters will be back. The first airplanes didn't fly either.)

The European Union has moved from a food deficit to a net surplus—at a price—by practicing intensive agriculture, using twice as much fertilizer per hectare as the United States and three times as much pesticides (OECD data). That intense agriculture, coupled with industrialization and high population densities, has led to serious air and water pollution. Nitrate loads in Dutch groundwater drinking supplies are four times the maximum U.S. standard.[119] The sparkling Adriatic I remember from 1965 is now afflicted with "The Brown Slime," a product presumably of heavy fertilizer use in the Po valley. European rivers are astonishingly loaded with pollutants. The nitrogen load in the Thames, for instance, is nearly four times that in the Delaware, one of the more polluted American rivers, 15 times that in the Yellow River in China and more than 600 times the load in the Nile. This is the price one pays for prosperity in a tight space. The European Union is trying, more seriously than the United States, to remedy its environmental problems, but it faces a major task.

Western and Southern European populations will probably stop growing in about 2010 (UN 1994 median projection) and by 2050 may be back down to the level of 1965. Northern Europe is expected to grow very slowly, reaching a plateau around 2040 at a level 6% above the present.[120] The dependency ratio for all those regions is around 50—one child or old person for two working-age people—which is very favorable indeed. It will begin to rise around 2020, reflecting the low recent levels of fertility. By midcentury, it may be as high as 80 in Southern Europe. This is high, but perhaps not altogether a problem for a continent suffering from unemployment rather than a shortage of labor.

One might expect that Europeans would greet these projections with pleasure and relief. As we shall see, Wassily Leontief (p. 84) is proving prescient. Unemployment is very close to a crisis level. The problem should slowly moderate as smaller age cohorts come on the job market. The coming reversal of population growth also promises to alleviate the problems of pollution, simplify the energy transition for Europe, take the pressure off agriculture and permit the reversion to more benign farming practices.

Are the Europeans happy? Hardly. Never underestimate the fear of "smallness" or of change. In a fascinating essay, several leading Germans including former Prime Minister Helmut Schmidt proposed ways to assure that "the Germans not die out."[121] They cited projections suggesting that, *at present fertility levels* (my emphasis) and without immigration, the German population will decline by about 1 million in 2000 and by 14 million to 70 million in 2030, and it will be an older population. Germans, they pointed out, will constitute just 0.6% of world population compared to 1.5% at present. (Does this imply that they see Germany in a population race with, say, Bangladesh?) They predicted that Germany will be less populous than Russia, Ukraine, France, England and perhaps Italy. (The idea of a "European" identity does not arise; they are thinking traditionally of Germans.)

The authors warned that relying on immigration to counter the decline would raise the danger of "marginalizing" ethnic Germans in a heterogeneous population. On the other hand, they noted correctly that it is very difficult to raise fertility by political means. Moreover, if they did succeed in arresting the population decline solely by sharply increasing the number of children, there would be a very difficult period immediately ahead when the working population would have to support a heavy load of both old and young people.

Perhaps most interesting, the authors warned about the "narrow shoulders of the younger generation" and the potential lack of labor, even though Europe's most pressing problem right now is unemployment. They warned that the "previous one-way street—ever less work for an ever higher standard of living—will come to an end not only on ecological, but also on demographic grounds." What they ignored is that the end of that road has already been reached—at a time when there is more than enough labor. Germany's dependency ratio is still just 45 per 100 working–age Germans, theoretically one of the world's most favorable ratios, and the number of unemployed has passed four million. The unemployment is not transitory; nearly half the unemployed have been out of work for a year or more. If the European Union stays on course, there are many unemployed throughout Europe who will seek work in Germany, if Germany ever again has a labor shortage. The dependency ratio is not Germany's problem.

Leontief's message has not yet reached those writers. Germany's well-being does not depend on the number of workers but on a solution of the endemic, worldwide problem of unemployment. Fewer workers would be a blessing.

The writers concluded, less than enthusiastically, by accepting the likelihood of some population decline, proposing "moderate" immigration starting with ethnic Germans and Eastern Europeans and trying to achieve a "measurably higher birthrate." To raise that rate, they said ". . . only when the reasonable material and spiritual needs of children in and out of family units have been satisfied should society be allowed to take up any further sociopolitical chores. The birth of children ought not to languish in Germany on grounds of material want. The public must recognize that the raising of children is absolutely its most vital assignment."

The authors are certainly right to call attention to the facts of demography and to demand that Germany formulate a population policy. Germany has had an exceptionally low fertility rate—1.5 or below—for nearly a generation, and it is now about 1.3. Population momentum works downward as well as upward. Ethnic Germans do indeed face gradual extinction if they do nothing.

However, the article doesn't answer the questions: How far down is tolerable? or perhaps desirable? The almost panicky urgency misses some key points: at 70 million, the population in 2030 would still be as large as in 1955, when West Germany was doing very well indeed. Population density would still be among the highest in the world at 200 per square kilometer (520 per square mile; compare the United States at 74 per square mile). The lower population would mean a 17% reduction in pollution and CO_2 emissions even without new conservation measures. The energy deficit and the need for food would decline similarly, which would offer an excellent chance to save the foreign exchange that buys energy and to ameliorate the intensive farming and levels of industrial activity that are polluting Europe's air and water.

Moreover, 1.3 is a remarkably low fertility. It may well rise somewhat, as has happened in the United States and Sweden. Such a gradual rise might be the best possible demographic future, leading to a smaller and more sustainable population but at a gentle rate of adjustment that does not produce panic about potential labor shortages.

Spain, Italy and Greece face demographic trends very similar to Germany's. For Austria and Japan, the issues are almost as sharply drawn. For most of the rest of Northern and Western Europe, fertility rates are not so far below replacement level. Those countries are headed toward a more gradual population turnaround.

Japan and Europe are two of the world's most densely populated regions. A gradual population decline strikes me as a very comfortable prospect. A smaller population will put them into a better balance with their resource base. They must indeed move toward higher birth rates, but they have a lot of time.

The eventual alternatives are higher fertility versus immigration and/or disappearance. I believe, however, that—despite Mr. Schmidt's fears—for at least the next two generations, population decline in Europe and Japan will put them in much better shape to deal with the problems they face than are the countries that are still growing.

The First Postwar Crisis of the Industrial World

The European labor situation reinforces that appraisal. Unemployment is approaching crisis levels through most of the industrial world, particularly in Europe. For the European Union it is 10.6%, over 20% in Spain and nearly that in Ireland. It is rising in Japan. Perhaps most ominous: it is much higher among the young, both in Europe and in the United States (chapter 10). That is a dangerous situation.

The Italian Minister of the Interior warned in 1993 of the danger of urban riots if the unemployment crisis continues. The U.S. Under Secretary of the Treasury has warned of a "huge increase in global unemployment" and said that the "administration's top economic priority [is] restoring job creation in industrial countries."[122] The Managing Director of the International Monetary Fund warned that high and rising unemployment is a "devastating trend," but he had nothing to offer but lower interest rates (which raise the specter of re-igniting inflation), lower budget deficits and trade liberalization (which I think is part of the problem, not a solution).[123] Europe and Japan, despite their comfortable and apparently insulated situations, are chin deep in the dangers of a world trading system that has lost its balance. We will turn our attention to that problem in the next chapter.

European leaders are trying hard to preserve their wage scales and welfare society. They are thus following the opposite tack from the United States, which has held unemployment down by allowing wages to slide and which has yet to institute the broad welfare programs that are general throughout west Europe. Neither "solution" has yet shown that it can work. Europe is stuck with its unemployment, and in the United States real wages are declining and income gaps are widening.

The "Group of Seven" industrial nations (G-7) met in Detroit in March 1994 and offered old nostrums for a new problem: economic growth, more trade and an admonition (expressed by U.S. Treasury Secretary Bentsen on behalf of the group) to "workers everywhere to recognize the benefits, and not just the pain, from rapidly advancing technology." They met again in April 1995 and returned to the same themes.

The world seems fixated on growth as the way out of all problems. West European GDP grew at an average rate of 2.6% per year through the '70s and '80s, but that was insufficient to avoid the crisis of unemployment in Europe (or the deterioration of wages in the United States). The editor of *Foreign Affairs* remarked to me that the growth rate must be raised. That is the conventional wisdom: "grow faster." Even the 2.6% growth rate means a doubling every 27 years. A rate of 3.5% would be a doubling every 20 years: 2, 4, 8, 16, 32-fold before the year 2100, 64, 128, . . . How many doublings do those conventional thinkers think that little peninsula can bear? How many can the United States bear? They are trying to solve their problems by intensifying an intolerable process.

You cannot grow out of your problems.

Perhaps Professor Bartlett was right to warn about 14 doublings (p. 55). Politicians seem to want to test the warning.

Others in Europe have proposed a deliberate erosion of Europe's high wages, benefits and health care and a shorter work week to spread the work. President Clinton took up the last idea. One part of his proposed Reemployment Act (now dead in the water) would have encouraged employers to shorten working hours rather than discharging workers. That is precisely contrary to most employers' present policy.

One or two leaders have suggested the unthinkable. A senior official of the OECD, refusing to be identified, said that protectionism, "silly" as it is, may be better than a "social explosion" in the industrial world.[124] Outgoing French President Mitterand, picking up a theme never very far from many French minds, called for higher European trade barriers against competition from low-wage countries.[125] The OECD Trade Union

Advisory Committee has called upon governments to slow or cushion the internationalization of trade and capital flows.[126]

■   ■   ■

Such drastic policy changes may be necessary. At least inferentially, those speakers identified the causes of the labor problem, but none of them suggested that world population growth and migration have any bearing on the problem—despite the fact that in almost all of the industrial world immigration has become an intense political issue. I suggest that they broaden their definition of the problem and their search for solutions.

The remarkable thing is that many governments in the industrial world still press for more children—meaning more unemployed youth later on—even though their citizens don't share their enthusiasm. Several countries offer bonuses, tax breaks and/or medals for child bearing (including the United States, inadvertently, in its income tax laws). French President Chirac recently invited a group of mothers of large families to the Presidential palace and assured them that "despite what some people say, it is by having more children that we will cut unemployment, and not the contrary."[127] Japan recently raised the bonuses, and both political parties in the United States seem intent on doing so. This may be a reflexive reaction in Europe, thinking in antique terms of relative national power, but we do not have that excuse.

CHAPTER 8

One World, Like It or Not

LET US SEE WHAT HAPPENS when we mix together a world facing such differing circumstances. I will start, as in chapter 7, with the poorest countries, move on to the needs of the emerging countries and their effects on the older industrial countries, relate these forces to the pursuit of common objectives such as protecting the environment, and wind up with a practical example of the impacts of population growth on "real world" decision making.

THE RIPPLE EFFECTS OF POPULATION GROWTH

The turmoil of population growth is concentrated in the poorer parts of the third world, but its effects will not stay there.

Pestilence

We have seen the prognosis of megacities and epidemics in the third world (chapter 1). On a single Earth, can the industrial world isolate itself? Epidemics spread; the industrial world has no basis to believe that they will stay conveniently in the poor countries. With modern communications, persons infected with disease can move halfway around the world, infecting others, before they themselves show symptoms. Plagues and disease, driven by poverty and crowding in the poorest countries, can migrate almost instantly to the prosperous ones. Cholera (from wind–blown fecal dust) appeared in Peru in 1991 and created a scare in Los Angeles. A cholera carrier was thought to have flown to Los Angeles from Peru, and I don't think the person was ever found.

Pathogens are a great leveler. Various strains of influenza, some of them new, start out each year from China. AIDS apparently started in Africa. The plague, cholera and new and resistant strains of tuberculosis have appeared or reappeared as diseases of crowding, filth and poverty in third world urban slums. They have reached the richer countries, including the United States.

Some of the threats we can handle with modern public health systems. Some we cannot. Health authorities have been trying for several years to tell us that there may be no defense against the multiplying strains of drug-resistant pathogens.[128] Self-interest alone dictates that we help the poor countries control their problems.

The Age of Migrations

There are nearly 3 billion working-age people in the third world, growing by 60 million each year, compared with a stable 800 million in the industrial countries. The graph shows the ongoing annual growth. The problem is not transitory. I have cited the ILO estimate of 30% unemployment, most of it in the third world, and the numbers have yet to peak.

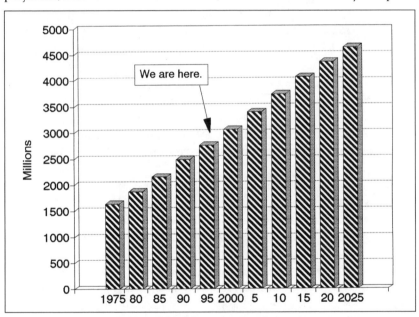

Fig. 8-1. Third World Working-Age Population Ages 15–64)
(*UN 1994 Middle Projection*)

Central America provides a good example. The economically active population is increasing at a rising arithmetic rate: from 3 million in 1950 to 6 million in 1975, with 12 million on the way by 2000 and 24 million projected for 2025.[129] This growth drives migration, and anybody in the world who plans to address unemployment without factoring in migratory pressures is bound to fail.

The first move is rural to urban, but we saw in chapter 1 that third world cities are exploding and offer little hope. Then the movement is across borders. The movement is with us already. The numbers are fuzzy; the FAO guesses the total at 100 million, including internal migrants.[130] They usually head for the rich countries but not always. Some go to less impoverished neighbors. Nigeria has from time to time evicted hundreds of thousands of illegal immigrants. So have Pakistan, Zaire, South Africa and Mexico, on varying scales. People are trying to move *into* Liberia and the Congo. That is a measure of the state of desperation.

As emerging nations begin to look like the older industrialized countries, they inherit some of the same problems. An official estimates that there are 230,000–300,000 Bangladeshis working in Malaysia, another 40,000 hiding in the jungles of the Kra Isthmus waiting for a chance to get in and more on the way.[131]

Long distance migration is a privilege of the more prosperous. Witness the African and Middle Eastern students driving taxis in American cities. These educated youths should be the next generation of leaders at home.

The pressures to move will stop only when population pressures stop, or when migration wipes out the disparities and misery is equalized around the world. The choice for the rich countries is either a fundamental reordering of priorities to prevent the migration or submission to it. It cannot be successfully absorbed by any country. The numbers are simply too great.

Immigration will probably affect Japan less than most industrial countries because of its homogeneity and its island location.

Europe has some serious problems. The control of migration will pose some wrenching decisions, given the flow of information and ease of international movement in this era and the existing European connections with Africa, Turkey and the remnants of the Soviet empire. Two-thirds of the European Union's 1.1 million population growth in 1994 was net migration.[132] The Union is in the process of lowering internal barriers to movement; immigration from outside the Union—most of it headed for the prosperous northern tier in Europe—threatens to halt that project.

The United States has the worst problems of all in addressing migration, as we shall see in Part III.

■ ■ ■

If poorer countries' populations were stabilized (or better, declining), they would face less daunting problems from the restless unemployed and from the destruction of their resources. The industrial world would face a less frightening prospect of migratory pressures from poor countries. It would be a more relaxed world.

The problem of excess labor is slow to respond to demographic change because a drop in job market entrants lags a drop in fertility by about two decades. Only in parts of east and southeast Asia and some Caribbean islands has fertility fallen enough to offer confidence that the pressure will disappear within the next generation.

TRADE

The complexities of a disparate world are played out most vividly in world trade and investment. The most dramatic issue in the next two decades will probably be a sellers' market in grain. Less obvious but equally critical, world trading arrangements profoundly affect income, employment and even political stability in both the third world and the industrial countries.

World Grain Trade

The emerging industrial countries cannot feed themselves. They will be competing for food in a world market that cannot accommodate them. In September 1995, the European Union temporarily suspended grain exports to keep a lid on its own bread prices during a poor harvest.[133] In the United States, the chairman of the House Agriculture Committee warned that farmers should be protected in light of "growing talk that exports be curbed due to short grain supplies."[134] That should ring alarm bells in food–importing nations.

In 1950, interregional grain trade essentially consisted of moderate exports (23 million tons) from North America supplying a similar deficit in Western Europe. Since then, North American exports have more than quadrupled, and West Europe has become a moderate exporter. However, Asia now runs a deficit of 80 million tons or more, Eastern Europe and the countries of the old USSR run a deficit of 35 million tons, Africa 31 million tons and Central America and the Caribbean 11 million tons, for a total annual grain deficit of about 157 million tons.[135] The older industrial countries are helping to feed Japan, the erstwhile USSR and the third world.

The third world is already a net importer of more than 100 million tons of cereals per year (including aid shipments). The Worldwatch Institute examined ten of the major food importers aside from China and concluded that their grain import needs will rise from 32 million tons in 1990 to 190 million tons in 2030.

Projections of the prospective increase in demand from China and the other major food importers thus range from about 150 to over 500 million tons in the next generation (chapter 7). Even omitting any growth in demand from smaller importers, the projections suggest total world demand will double or quadruple. The world's exporters are most unlikely to be able to supply even a doubled demand. In fact, the plateau of yields (chapter 2) may reduce exports. The United States is the residual world supplier, but it is faltering (chapters 2 and 9), and U.S. population growth may reduce our exportable surplus.

Grain prices are rising, and nobody knows where the competition will drive them. Food is the first priority for any nation. Deficit nations will push their exports to buy grain. Higher grain prices and rising imports will almost certainly lead the old industrial countries to raise trade barriers, which in turn will leave those food-short new players without recourse.

Meanwhile, the exporters' balance of payments problems are already leading them to sell their grain rather than give it away, and the poorest nations, such as most of sub-Saharan Africa and Bangladesh, are increasingly frozen out of the market even as food aid declines. World grain surpluses are already declining to something approaching a crisis level (p. 17). The FAO says that world food security in 1996/97 will depend "critically" on a substantial production increase in 1996.[136]

Japan is heavily dependent on imports for food as well as energy. At nearly 30 million tons per year, it is by far the largest importer of grains. Two trends will shape its future reliance on imports. Population growth is expected to turn around by 2010, gradually reducing food needs. On the other hand, arable acreage has declined 17% since 1970. If a smaller population makes it possible to stop diverting acreage from agriculture to urban uses, Japan's need for food imports should stabilize—but not disappear. The United States' continued ability and willingness to export grain may be more important to Japan than our military alliance.

West Europe is relatively secure. The forthcoming reversal of indigenous population growth means that it should be able to continue to feed itself—unless barriers to immigration really collapse.

■　■　■

Because there has been plenty of grain in world trade for years, we tend to forget its importance. Nobody can predict what will happen as powerful nations run up against their peoples' inability to get enough food. The population connection here is pretty clear: the smaller the population of importing states, the better they can avoid a level of dependency that is fated to be disastrous; the smaller the populations in exporting countries, the more will be available for export.

MNCs: the New Players

I have observed (chapter 5) that cheap fuel and successive waves of technological change have led to unprecedented levels of world trade and economic interdependence. A new economic creature has evolved to exploit and manipulate the new opportunities: the multinational corporation, or MNC (also called the transnational corporation or TNC). Large corporations used to be identified as "British" or "Dutch" or "American" or some particular country's. Those identifications are less and less relevant, as the average person discovers when his or her new "American" car arrives from Canada. A "Japanese" car may come from Japan, Korea or a factory in the American heartland.

The suprasovereign MNCs owe allegiance to no sovereign; they are themselves stronger than many of the countries in which they operate. The ILO has observed that

> The world of the 1990s is very different from that of the 1950s, when issues of employment, income inequality and poverty in developing countries first attracted the attention of policy-makers. The world at mid-century was not closely integrated.
>
> Today globalization has triumphed. The world economy is more closely integrated than ever before: the market is rapidly superseding government controls and planning as a mechanism for allocating resources; . . .
>
> Globalization has had two consequences . . . First, it has created opportunities for material betterment . . .
>
> A second consequence of globalization is that it has weakened the ability of individual States to manage their economies. (World Employment, 1995, p. 68)

A UN study in 1983 found that 56% of the net earnings of 380 of the largest industrial corporations came from foreign affiliates.[137] Private foreign direct investments averaged $150 billion yearly from 1986 to 1992. MNCs' total employment has been stable, but their employment in foreign countries rose from 22 to 29 million from 1985 to 1992 and is now larger than their "home" employment. They are indeed "exporting jobs."

I do not see them as a conspiracy but as powerful entities pursuing their own interests. The MNCs see the world as a single market in which they move production, management and capital to maximize their profits. They argue that national economic boundaries have become irrelevant and that the future lies in a world trading system. They seek the cheapest good and trainable labor in places where there is adequate economic infrastructure such as roads and harbors and the future is not too uncertain. They are not responsible for the well-being of labor as governments are (or should be). They are a pervasive, quiet presence with a massive investment in open trade and capital movements.

The rise of the multinationals, pursuing their own interests, introduces powerful participants in world trade whose interest in the general well-being is untested and certainly secondary to their "bottom line." It injects an element that nation-states have not yet learned to handle. The MNCs' reach is awesome. Witness the 1994 collapse of the Mexican peso. The most immediate losers were the international bankers holding loans to Mexico. The Clinton administration had promoted the North American Free Trade Agreement (NAFTA), with the encouragement of the MNCs and over the bitter opposition of Clinton's labor allies. It was supposed to prevent precisely such a collapse. Its credibility was at stake. When Congress refused to go along with a $40 billion loan guarantee to Mexico, Clinton bypassed Congress and guaranteed $20 billion in loans from a fund that had been created to protect the U.S. dollar—and there were almost no complaints from Congress. International bankers have their little ways. The peso recovered somewhat and then collapsed again in late 1995, but the bankers had their guarantees, and Mexico did not default.

My point is that trade policies, investment and the world banking system respond to the wishes of the powerful, not necessarily to the needs of peoples. Hardly a novel thought. The MNCs are a loose cannon on the deck at a very difficult time.

Some liberals complain that the U.S. Government has allowed the multiplication of cattle ranches in Central America that produce cheap beef for the American market and in the process displace peasants and force them into cities. The critics have a point; the process is comparable to the "enclosure movement" two centuries ago. (That movement drove peasants off the farms of Scotland and northern England and replaced them with sheep after the invention of the "spinning Jenny" multiplied the industrial demand for wool, to the enrichment of English landowners, textile mill owners, and traders.) The liberals' target should be MNCs and free trade. The Central American governments are too weak and perhaps too compromised to resist the ranchers, and the U.S. Government, wedded to free trade, is in no position to set rules banning hamburger imports, even though the displaced peasants may wind up in the United States.

Trade and the Environment

Free trade is no friend of environmentalism. The MNCs are unlikely to be environmentalists, and the power of the free trade alliance is a threat to national environmental legislation. The General Agreement on Tariffs and Trade (GATT) in its day had regulatory powers to punish actions taken "in restraint of trade," and it did so. It found against the United States when we tried to apply U.S. laws to tuna imports caught in violation of U.S. standards of dolphin protection. The new International Trade Organization (ITO) has more extensive power than GATT did to enforce such decisions. It has before it a Venezuelan complaint that U.S. gasoline standards, intended to make gasoline less volatile, are in restraint of trade. Several other countries expressed support for the Venezuelan complaint.[138] I am in no position to make a judgment on the merits, but I would point out that this is an environmental issue being adjudicated by people whose interest is free trade, not the environment. The European Community Court of Justice, largely a trade management body, forbade Denmark from enforcing a Danish law requiring sellers of foreign as well as domestic beer and soft drinks to recycle the containers.

Free Trade and the Poor

The United States is pressing other countries to dismantle impediments to free movement of goods, services, capital and management around the world. This movement is enshrined in trading arrangements such as NAFTA and the ITO, which are powerful instruments promoting free or freer trade, without much regard to the social or environmental consequences. Governments are signing on to declarations of intent to move to free trade, such as a statement by the Asia Pacific Economic Cooperation (APEC) forum in 1994 calling for free trade among all its members by 2020.

To a considerable degree, these declarations are meant to placate the United States. Malaysia, China and South Korea made clear they do not see the statements as binding. Observers at a recent APEC meeting remarked that the Asian participants are well aware of the Mexican experience with NAFTA and determined to keep control of investment and trade in their own countries.[139]

In this jungle of acronyms lurk some of the toughest and least recognized questions facing the world, and they are made tougher by population growth.

Economists have a convenient straw man called "economic man," motivated only by economic considerations, completely flexible and substitutable. The real world doesn't work that way. When displaced, peasants do not simply become, say, computer technicians. Technological change may be beneficial over time, but the process of introducing it is usually disruptive. The process used to take generations. The MNCs can now apply a new technology anywhere in the world almost instantly, and they do not necessarily think of the consequences.

The revolt in Chiapas, a poor and remote province of Mexico, offers a lesson about free trade and the third world. It began, symbolically, on January 1, 1994, when NAFTA came into effect. The peasants feared that the competition from efficient American corn producers would wipe out their cash crop, deprive them of a livelihood and force them to follow other peasants into the cities.

For the poor countries, the impacts of free trade, both on traditional agriculture and on urban workers, may be disastrous. They cannot simply shed one system and adopt another. The difficulties that the countries of the erstwhile USSR are having in adopting market economies are a warning. Open worldwide competition will generate enormous social strains within the third world. This is true even of a country such as China that is entering the modern world, albeit very unevenly. Domestic competition and the progressive removal of trade barriers imperil shakier industries and threaten massive layoffs. In China, the issue is already joined, and the government is torn between the search for efficiency and the need to preserve jobs. I suspect that China will have no qualms about controlling international trade to prevent imports from precipitating factory closures.

If there were not so many poor in the third world, the process of modernization could proceed with less suffering. Modernizing countries would eventually benefit from their higher productivity. However, the poor are there in growing numbers, and the free trade prescription is a merciless one.

Trade, Wages and Industrial Countries

Conventional economists (particularly in the United States) would have us believe that unrestrained free trade is a virtue and that any controls are a sin. MNCs, seeking cheap labor and expanding markets, agree. Governments tend to listen to them.

To placate the conventional, let me reiterate that trade has a valuable role (p. 87). It has permitted a degree of specialization that has enabled a larger population to live better than ever before through the operation of "comparative advantage." It lowers prices for the benefit of consumers and creates some jobs (though not so many as proponents claim) and business opportunities for exporters. Politically, it may give countries a stake in the stability of their suppliers and markets and in a stable world—up to a point. However, the issue is, How much trade, and under what conditions?

The doctrinaire attachment to free trade can have dangerous consequences. For the richer nations, the immediate problem is the impact on employment and labor. With high wages and labor standards, they face a fundamental question that free traders have yet to answer:

> *In a world with free movement of capital, technology, management, marketing systems and goods, can the prosperous countries preserve their wage differentials?*

MNCs and the free trade system itself conspire to move production to the cheapest place and produce goods to be shipped back to the industrial world markets. In industries where labor is a significant fraction of total costs, the work migrates to places with lower labor costs.

The proponents of free trade are either (a) unconcerned about the labor impacts, or (b) willing to suspend their belief in the fundamental economic law of supply and demand in the sole case of labor, or (c) both.

The free traders have managed to establish the myth that all true economists are free traders. Not so. Herman Daly has found some pungent quotes from Adam Smith, David Ricardo and John Maynard Keynes to the contrary. Consider this one from Keynes:

> . . . let goods be homespun whenever it is reasonably and conveniently possible; and, above all, let finance be primarily national.[140]

Can the rich preserve a prosperous island in an impoverished world? Militarily, probably so. Economically, they cannot reconcile differential living standards and free trade. That is a tough one to handle because the need to protect domestic labor conflicts with the wish to accommodate emerging nations' need to export.

That question in turn raises another one: *Should* the industrial nations try to preserve the differentials? For me, at least, the answer is straightforward: yes. In a world with an expanding and (for practical purposes) unlimited labor force, the alternative is to drive the whole world progressively toward subsistence–level wages. In a world without fertility differentials and without an explosion of job seekers in the third world, one might hope that wages could indeed converge and at a higher level rather than a lower one. We don't live in that world. That is the message that the free traders should be hearing from the demographers. They don't hear it.

The Old Rich versus the Emerging Nations

We in the United States fear Japan's competitive prowess, but Japan is far from immune to the problem of low labor–cost competition. It has been losing export markets for heavy goods and consumer durables to lower wage countries for years. Korean automobile exports rose to over 1 million in 1995, an increase of 44% in one year, while Japan's auto exports stagnate.

Right now, Japan is still doing better than most industrial nations, despite its jitters about the end of its economic "miracle." Its unemployment rate is only about 3%, and it has less of the disguised unemployment rampant in the United States (see chapter 11), though it too is having trouble with unemployment among the young. It manages to contain the problem of labor displacement through managed trade—its protestations to the contrary notwithstanding. Japan will probably be able to handle the competition better than most industrial nations, given its homogeneity and discipline, but managed imports raise the fear of retaliation, and Japan must export enough to pay for the food, energy and raw materials that constitute half its imports.

After Japan, will Korea in turn fear low-wage competition as its labor force stabilizes and wages rise? Each nation has its day in the sun and then is endangered by exporting nations with lower wages, organized and managed by MNCs or (if lucky) by forward-looking governments. The free-trading system sets in motion a cascade of rivalries. At the present stage, the old industrial world is in competition with emerging low-cost producers. As they in turn prosper, they will go on the defensive against lower cost producers. Everybody can play. Even Bangladesh, organized by U.S. importers, is shipping piece goods to the United States.

What will the industrial world do? The "school" solution is economic growth. It is the wrong solution. It won't work indefinitely because of the inescapable limits to growth described in Part I above. And it won't work now with free trade.

Do we reconcile ourselves to meeting the competition and letting the poor countries' labor prices dictate wages here? We know the world labor force will rise; the work force entrants of a generation hence have already been born.

When these questions arise in connection with U.S. trade policy, there is a standard answer. Look, say the conventional, our balance of payments problems are with higher labor cost countries, not the poor ones. Look at the way our exports to Mexico have risen! Look at the jobs those exports create!

Look again. We generally ran a merchandise trade surplus with Hong Kong and Japan until the early 1960s, with Taiwan through the late '60s, with South Korea until the mid-'70s, with Singapore until the early '80s and with China through the mid-'80s. The jobs went overseas with the movement of industries to those countries. NAFTA was justified in part on the grounds it will help Mexico modernize. If it eventually works, be prepared for the competition.

The more advanced economies are forced increasingly to rely on high-tech exports. The U.S. International Trade Commission (USITC) study that justified NAFTA, insofar as it dealt with U.S. labor impacts at all, assumed that the United States would keep the skilled jobs and Mexico the unskilled ones. Our experience should teach us the falsity of that patronizing assumption. The United States may well keep a leading role in cutting-edge technology (if our universities don't go the way of our primary and secondary schools), but except for aerospace, the big net foreign exchange earnings are not in high-tech products. To make automobiles or clothes or appliances, the MNCs have every reason to produce where labor is cheap and sell here. China, for instance, is already upgrading its exports; Chinese machinery and electronics exports jumped 60% in the first nine months of 1995, compared to the same period in 1994, to become the biggest export category; fax machine exports to the United States rose 30-fold.[141]

The hope—in Europe, Japan and the United States—of preserving a niche at the attenuated top of the world market is a will-o'-the-wisp.

The USITC study for NAFTA (above) was frank in admitting that, while differences in population growth rates might sustain migratory pressures into the United States, affecting U.S. wages, "this possibility was not examined separately."[142] It thus missed the central issue. Free trade may work in a stable world. It does not work in a world of enormous wage differentials that are sustained by continuing third world population growth. The economists' world model is a stable one; it does not accommodate demographic change.

The drive toward free trade poses another danger that is not in the economic models: real or potential instability (chapter 5). Is the world so stable that we can afford the level of interdependence the theoreticians would urge on us? Unstable energy and food supplies. The sense of personal helplessness when people depend not on adjacent woodlots and fields but upon developments half a world away. At this point, the dangers of interdependence (chapter 5) and the effects on labor reinforce each other and argue against free trade.

One issue stands out: in every current and recent major multilateral trade negotiation, agriculture has been a sticking point as other countries resist efforts by the major commercial agricultural producers—the United States, Canada and Australia—to reduce or eliminate tariffs on cereals. There is a reason for the others' concerns. I have mentioned the Chiapas revolt in Mexico (p. 140). The Mexican Government apparently had not thought out the consequences of NAFTA for unemployment in their country, nor had we. Other countries, the ones that resist free trade in basic foodstuffs, are apparently more sensitive to that issue than Mexico was. They may also be protecting a portion of their food supply from the vagaries of the international market. Efficiency is not everything.

These arguments all point to the likelihood of a protectionist future rather than freer trade. Japan and Europe are adept at managing their trade with nontariff barriers, and neither has the emotional commitment to free trade that the United States does.

The result? The United States, which has already provided the market on which the countries of East Asia have taken off, will be under ever more intense pressure to maintain an open import policy. When that fails—because we cannot afford it and our purchasing power is falling—we will be enormously tempted to follow the European and Japanese examples. Then will come the effort to divide the world into trading zones.

What happens as a growing part of the populace is unemployed and growing poorer? The threat of social breakdown. The industrial nations have strong reasons for wanting to avoid the competition from educated and/or trainable nationals from poorer countries. The NAFTA and ITO tariff proposals are scheduled to be phased in over a period of years. It does not require prescience to predict that intense pressures will arise to halt the process so as to limit the damage.

■ ■ ■

I will argue the need for a changed U.S. trade policy in chapter 12. For the present, let me underscore the population connection I make on page 147. Differential fertility is the central impediment to a successful pattern of international trade. Third world population growth is the principal agent that drives the tensions between the old rich and the emerging countries. If somehow it should slow and stop, the world could look forward toward a time of adjustment when the emergence of third world nations would not pose such a threat to the environment, when wages would begin to converge, when international trade would not endanger industrial countries' labor standards, and "comparative advantage" could play a benign role in raising the productivity of both sides.

THE ENVIRONMENT

We are one, essentially destructive, species on one fairly small planet. We have a shared interest in preserving the habitability of that planet. As companion voyagers in the hugeness of the universe, we must cooperate to survive.

I am arguing, perhaps perversely, for a sense of interdependence in the stewardship of a shared Earth, even as I argue against too high a level of economic interdependence. Perhaps the best short answer is that we cannot escape environmental interdependence.

We must have learned something from the photographs of Earth from space. They should lead us beyond the fierce tribalism that has marked most of the human experience. International relations are not a zero–sum game as they have often been seen in traditional diplomacy. Your loss is not necessarily my gain. We have more to gain from cooperation than from competition.

Many of the major environmental and resource problems have international ramifications even if they must be attacked country by country: climate, acid precipitation, sustained food production, energy, forests, fish stocks, fresh water.

We—the nations of the world as surrogates for a common humanity—must cooperate to preserve the environment if we can and to manage the more mundane issues of living together: resource disputes, migration, health, trade and finance.

We have looked briefly at the terrifying implications if even a portion of people in the poor nations should begin to consume as the industrial nations do now. The prospect of a multifold increase in world GNP (p. 123) is intolerable, even if it were accompanied by draconian technical measures to limit pollution. Even the present world population would destroy the environment if everybody lived at the level of the rich countries. The combination of projected population growth and high living standards is out of the question.

Environmentally, as the old rich are augmented by the new rich and the numbers of high consumers rise, some combination of fewer people and a changed lifestyle will become imperative as a matter of survival. Not an easy task.

The Pursuit of Wealth

When it comes to practical policies, the sense of belonging to a larger community will not come easily. The emerging countries and the rich in even the poorest countries are telling us that they plan to be as rich as they can be and that the environment is a very distant priority. It is going to be difficult to dissuade them—particularly since we already "have ours."

Supported by a seemingly limitless abundance of fossil energy, first the United States, then Europe and Japan and now much of East Asia have

145

been swept up in a remarkable spurt of growth. Growth seemed to work before, and the natural response to challenges is to go back to that solution. If we are running out of oil, create incentives so the industry will find more. If people are hungry, they need a green revolution.

The apparent success of industrialization in the West has bequeathed a dubious legacy to the less prosperous: the idea that unlimited growth of individual consumption is possible and desirable. This idea is really quite novel in human history; mankind has heretofore been constrained by hunger, accident and disease—"fate." Those limits have not disappeared but were temporarily expanded by fossil fuels and technological change.

GNP (or GDP for the more sophisticated) is pursued as a mystical goal, even though it may have little relationship to well-being. It may be a measure of resource destruction rather than success. In the industrial world, GNP growth may measure such modern amenities as traffic jams, industrial pollution and the subsequent expenditures on trying to control the pollution. That is the direction in which the emerging nations are rapidly moving, but observing our problems does not seem to change their goals.

Some economists recognize the misleading nature of standard GNP calculations and have proposed new ones. In Indonesia, Robert Repetto found that GNP growth dropped sharply when the destruction of forests was counted as a capital loss. Herman Daly and others have proposed variants of "green" GNP calculations.[143]

Population, Consumption and the "Right to Pollute"

The present concern about climate change provides a fascinating insight into the problems of doing anything cooperatively about our environmental future.

The Berlin Conference. Climate was a central issue at the 1992 UN Conference on the Environment and Development (UNCED) in Rio de Janeiro. As a result, the industrial nations pledged to hold CO_2 emissions by 2000 to the 1990 level. In itself, that is far from sufficient to keep the amount in the atmosphere from continuing to rise (see chapter 4), but we are failing in even that modest commitment. U.S. emissions in 1993 were 2.6% above those in 1990, and the Department of Energy expects them to be 10% higher in 2000. It expects emissions in the European Union to be up by 16%.

When asked by reporters why the United States is falling behind its pledge, Department of State Under Secretary Timothy Wirth said the problem is that "oil prices dropped and the economy recovered more rapidly than expected." This illustrates the industrial countries' dilemma very succinctly. Their only cure for unemployment is growth, but growth compounds the difficulty of pursuing environmental targets.

In March/April 1995 there was a major international conference in Berlin to follow up on the pledges made at the UNCED. Before and during the conference, the players staked out their positions. The President of the Maldives pointed out that a three-meter rise in sea level would wipe out his entire country, and the Association of Small Island States (AOSIS) led a coalition of 60 third world countries calling for a reduction of CO_2 emissions by the industrial states to 80% of 1990 levels by 2005. (Still, I would note, far from enough to control the problem.) The Arabs, on the other hand, tried to block any proposal that would depress oil sales. A "senior Qatari official" told reporters that "if there is any threat to the environment, developed countries are to blame . . . because they have indulged in excessive consumption and materialism." He went on to blame coal and nuclear power for any such threats—but not oil.

The Indian Environmental Minister proposed that the industrial nations pay $100 billion per year as "environmental rent" to the less developed countries but insisted that the third world would undertake no obligation to limit its emissions, which were "bound to go up." He said it was up to the industrial countries to repair the problem because "we have not contributed [sic] to this global mess."

China already generates over 10% of world greenhouse gas emissions, behind only Russia and the United States, and its share is rising (chapter 7). Total third world emissions are expected to pass the OECD countries' before 2020. The Chinese National Environmental Protection Agency director, somewhat less truculent than the Indian, said that China hopes to address emissions by washing its coal but that it is an expensive process, and "economic growth is our priority." Together, China and India led the third world in resisting any commitments. They do not profit over the long term by claiming the right to pollute, but it is a reflexive reaction.

There was some disagreement within the industrial world but, led by the United States, Japan, Australia and the United Kingdom, it refused to make any binding commitments. The one concrete suggestion to come out of the conference was a proposal by Timothy Wirth, the U.S. delegate, that the developed countries help to finance low–emission technologies in the third world and be given "pollution credits" in return. Third world speakers showed some interest but demanded a delay in instituting such credits.

In the end, no specific commitments were reflected in the conference communique. The industrial nations accepted that "the share of global emissions originating in developing countries will grow to meet their social and economic needs." They agreed to set "quantitative objectives" for specific time frames of 2005, 2010 and 2020, covering their own emissions by source and "removals by sinks" (e.g., forests). They would not demand "any new commitments" from the developing countries.

The observer from the Global Climate Coalition, an organization of U.S. energy and industrial interests, told reporters that the United States had gone too far in agreeing to specific timetables. On the other hand, the President of the Reinsurance Association of America was also there. He was there to urge action, not delay. Hurricane Andrew did "over $17 billion" of damage in Florida in 1992; such a hurricane hitting Miami or New York could, he said, cause $50 billion in claims, or nearly one-third of the capital base of the entire American insurance industry. He described global warming as a major economic threat. Hurricane Andrew drove insurance rates in Florida up by one-fourth; reinsurance premiums went up four or five times, and reinsurers' liability is now capped. Soon, he said, it may be impossible for people in such zones to get insurance at all.[144]

Tugged by these conflicting interests, the conferees agreed to meet again in 1997 but on little else.[145]

The Third World and the IPCC. Third world scientists and representatives in the IPCC made clear that they would not agree to proposals to counteract global warming by stopping tropical deforestation until the industrial world brought its greenhouse emissions under control. We have been warned.

All the more reason for a population policy. If you cannot conserve enough, the problem would be smaller if a lower population were setting demand.

The Universal Stumbling Block. That feeling of a "competitive right to pollute" may turn out to be the rock on which international environmental cooperation founders. The third world is not alone in assuming, at least implicitly, the "right to pollute." We in the industrial world create the most damage with our combination of population size and consumption levels. Most of the world's population growth is occurring elsewhere. We say complacently that "they" have the problem. Not so. We in the industrial world have the problem too. There is neither a "right to grow" nor a "right to consume" nor a "right to pollute" if we are serious about preserving a habitable Earth. If only in self-interest, the rich nations should look at what they can do to reduce their impact.

We can and should attack the rich countries' environmental impact ("I") through conservation ("C") and technological fixes ("T"), but we should also look at population policy ("P"). Except for the United States, most industrial countries have roughly stable populations, but those populations are very dense. Western Europe is still the world's most densely populated subregion. Perhaps it is too crowded. If the industrial world hopes to maintain anything approaching the consumption habits to which we have become accustomed, we must reduce the environmental impact by encouraging the coming downward drift of population in Europe and Japan (chapter 7) and urging it on the United States.

Policies in the rich countries to lower our own populations would also help to legitimize population programs elsewhere by demonstrating that population assistance to third world countries is not, as some of them claim, "genocide."

Envy and Discontent: Taming the Rich

The problem of cooperation is made more difficult when the poor watch the poverty gap widening. Can the rich consume and expect the rest to stay poor and compliant? A disengagement between the rich and the desperate is not a very auspicious starting point for cooperation in protecting a shared Earth.

In an earlier era, wide disparities and shifting fortunes among countries were the norm, and they did not pose much of a problem. That insularity has disappeared, driven by the awareness of opportunities in other lands and the ability to move there.

The industrial world is just beginning to emerge from the "Age of Exuberance." We who were lucky must remember that that age was not universal. Most people in the poorer countries have never experienced it but are painfully aware of the pleasures of consumerism, because there are television sets even in remote villages. There is a complex mixture out there of envy and resentment: the desire to have what we have and anger at the rich countries for having it. A deliberate downward leveling is a difficult prescription, but the rich must learn to cooperate, even as most of the people on Earth have to adjust to diminished dreams.

It will be an epochally difficult task to avoid an adversarial stance, maintain a humane view of each other and see the need for cooperation when nations are in a sharpening competition for food and energy and when they face a tightening future. Less flaunting of the disparities might make cooperation easier.

I am speaking as much of individual consumption as of national income disparities. A rich and consuming class has been growing even in nations such as India and China where the majority are still very poor. A dramatic concentration of wealth is taking place within the United States. Such disparities add to the tensions as the less prosperous begin to lower their own expectations.

It may be politically difficult, but leadership and tax policies can generate inducements to a less wasteful lifestyle. Aside from the direct and immediate gains in pollution control, such a change would provide examples—to the world and to the less fortunate within countries—of environmentally benign behavior and of the willingness to address consumption.

Population and Third World Leaders: the Beginning of Wisdom

Most third world leaders crossed the bridge some years ago and recognize—even though the United States does not—that population growth may impede their dreams of prosperity. In October 1995, Indonesia's President Soeharto presented a "Statement on Population Stabilization by World Leaders" to the UN Secretary General. The Statement was signed by 75 heads of government of countries totaling 3.9 billion people. The statement said, in part:

> Degradation of the world's environment, income inequality, and the potential for conflict exist today because of rapid population growth, among other factors. If this unprecedented population growth continues, future generations of children will not have adequate food, housing, health services, education, earth resources, and employment opportunities.
>
> We believe that the time has come now to recognize the worldwide necessity to achieve population stabilization and for each country to adopt the necessary policies and programs . . .
>
> . . . we earnestly hope that leaders around the world will share our views and join with us in this great undertaking . . .

Unfortunately, the industrial world has not accepted the invitation. The only industrial countries among the signatories are Austria, Iceland and Japan, even though the prime movers in the declaration were two U.S. population organizations.[146]

It is clear that those third world leaders would welcome support in their efforts to address population growth. The industrial countries would benefit if they offered more of it.

THE "REAL WORLD": THE EXAMPLE OF WATER

This chapter has tended to focus on the long term. Our immediate "real, practical world" experience is also permeated with the population issue. What nations want to do may be impossible because of population dynamics, or what they do may worsen population growth and its consequences. Most governments and most people do not recognize it, but they have an investment in the population issue even if they have chosen to dismiss it as too arcane and too far off for their attention.

By way of illustration, let me cite just one of the issues around us that is being driven by a population growth to which our policy makers seem to be blind.

No single foreign policy issue—not even Vietnam—has absorbed so much of U.S. Presidents' time and energy in the past 30 years as the effort to create a stable peace between Israel and its neighbors, in the face of Arab opposition to Zionism dating back to the 1920s. The United States has undertaken the responsibility to keep Israel afloat in the meantime, and it has cost us more than $50 billion to do so. That aid fills over half of Israel's foreign trade deficit, but the questions are never asked, Is Israel viable and is stability attainable in that region? The answer may lie in intertwined issues of population and water as much as in Arab intentions. The U.S. Department of State is aware of the water issue but has not factored population growth into U.S. policy, even though the Secretary of State recently mentioned it.[147]

The population of Israel has grown from 1.3 million since its founding to 5.6 million. It is expected to rise one-third to 7.5 million by 2020 (UN 1994 median projection; it will be more if recent immigration levels continue). Of the present total, some 18% are Palestinians, whose fertility is almost twice that of the Jews—who themselves have more children than women in other modern societies. There are 1.5 million Arabs in the occupied Left Bank, with a fertility rate of 5.6 children and a population doubling time of less than 25 years. In the Gaza Strip, there are 750,000 Palestinians crowded into 350 square kilometers, with an astonishing fertility rate of 7.1 children, comparable with the highest rates in Africa. At that rate, the population will double in 15 years.[148]

Those growth rates put relentless pressure on jobs, but the critical problem may be water. Arable land is not growing, and in Israel there are already 11 persons per hectare. Israel depends upon imports for more than half its food and nearly 90% of its cereals. If peace could be achieved, the Israelis might be able to cut expenditures and earn enough by exporting goods and services to cover their food deficit, but I doubt the Arabs would fare so well. The Arabs already claim that Israel has diverted water from the part of Lebanon it controls and from West Bank Arabs to supply Israel and its West Bank settlers. They charge Israel with allocating more than four times as much water per Israeli as per Palestinian—hardly a situation that promotes enduring peace. In the Gaza Strip, they claim, water is being pumped at more than twice the replenishment rate, and there is saline intrusion as a result.[149]

Water is the coming issue in North Africa and the Middle East. In 1960, total annual availability in the region was about three times the theoretical requirement (conventionally put at 1000 cubic meters per capita per year). By 1990, it had dropped to 1436 cubic meters; by 2025 it will be only 667 cubic meters. Jordan will have 91 cubic meters, Syria 161, and Israel 311.[150] There have been multilateral talks about regional water problems, but they have produced more invective than results. The Arabs charge that Israel is holding onto its land conquests in order to safeguard its water supply, and a 1991 Israeli study confirmed that those water resources have become more important to Israel because of population growth.[151] Meanwhile, the Palestinians insist that any permanent agreement must provide for their recovery of water rights from Israel.

Water supplies are roughly constant. Population growth drives the worsening numbers. Desalinization costs approximately $3 to $15 per cubic meter.[152] Costly as it is, it will be needed to accommodate a problem so immediate that it cannot wait for population policy alone as an answer, even if population policies were adopted—and Egypt is the only government in that contentious area that deliberately seeks to encourage lower fertility. But desalinization for irrigation produces extraordinarily expensive crops, particularly if the world price of energy rises. The water issue may force some fundamental reordering of priorities in the Middle East, and that is not all bad.

U.S. policy depends on Egypt as the one reliable intermediary between Israel and the Arab nations, but Egypt's situation is at least as desperate. Its population has trebled since 1950 to 63 million. It will probably pass 92 million by 2020, and there is no clear prospect as to how those 92 million, and others that will follow, will support themselves. Egypt is well into the nitrogen trap; it uses nearly four times as much fertilizer per hectare as the United States does and manages to meet about 60% of its needs for grain. (Egypt used to be a food exporter.) It uses the entire flow of the Nile and is embroiled with upriver countries that are threatening to take more water and reduce the flow. It must already support 20 persons per hectare of land, more than any other agricultural economy, even Bangladesh. Overall, exports cover about half its total imports. Remittances from Egyptians overseas cover part of that gap, and U.S. aid has filled much of the rest, but the question arises, How long? Egypt is already troubled by Muslim fundamentalist terrorism, and its stability and continued moderation are far from assured.

U.S. economic aid to Israel is simply handed over, not programmed. Population program assistance to Egypt has been a negligible fraction of total aid. Particularly given the competitive attitude that encourages high fertility in the region, one would assume that, at the very least, the United States would be mobilizing every possible effort to seek a solution to the twin problems of population growth and water shortage or, if they are insoluble, preparing a policy to accommodate to the rising tensions.

I may be unfair to select an issue that is driven as much by domestic political realities as by foreign policy. The problem for our foreign policy has been the effort to balance our strategic interests in the oil-rich Middle East with the tremendous domestic power of the Israeli lobby. Nevertheless, I am still entitled to ask whether we would not be more likely to develop a sustainable foreign policy in the Middle East if we incorporated population growth and its consequences in our policy making, and not just domestic politics versus oil.

Let me urge you to try an experiment: take any major policy issue that troubles you, and ask, What is happening to population? How does that affect the issue? There may be a few for which no connection is apparent, but not many. We will explore some of those connections in later chapters. Try the same experiment again at the end of the book.

PART III

The United States: Issues, Temporizations, Solutions

LET US TURN NOW to the ways in which population growth, here and abroad, drives the U.S. future. First, I will look at U.S. population growth. Then I will relate that growth to the pursuit of national goals. Population growth stands between us and the pursuit of some of them. Conversely, pursuit of those goals can have demographic results we didn't expect and don't want. Most of our political leaders try to avoid making those connections. I will suggest some ways of ameliorating the problems, but temporizing on the population issue makes our tasks much more difficult.

I will suggest how the nation could come to grips with population. In theory, it should be fairly easy. In practice, given the politics and the emotions involved, it will be very difficult. Since U.S. population growth is presently driven mostly by immigration, the issue is squarely posed: do we try to protect a living standard higher than the poor countries have? Can we? Where does our principal obligation lie?

I will offer some thoughts as to how the nation might be persuaded to think about population growth and conclude with a look into a future very dimly seen.

CHAPTER 9

U.S. Population Growth: An Accidental Future

THE UNITED STATES IS ALONE among the large industrial nations in its continued population growth, something that has happened by accident rather than design but has tremendous ramifications.

OUR DEMOGRAPHIC HISTORY

The nation's population more than trebled from 76 million in 1900 to nearly 270 million now. About 43% of that growth consisted of post–1900 immigrants and their descendants.

It is useful to graph the numbers, starting with our first census in 1790.

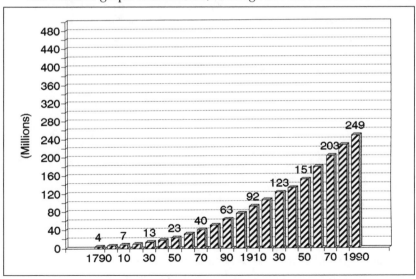

Fig. 9-1. U.S. Population, 1790–1990
(*U.S. Census Data*)

That growth is taking place in a space that no longer grows.

COMING CHANGE

A conservative projection leads to a population of 397 million in 2050 and 492 million in 2100 (figure 9–2).[153] About 91% of the growth after 2000 will be post-2000 immigrants and their descendants. (The lower bars show where we would go without any migration after 2000.) Immigration is the driving force in a remarkable and continuing population surge.

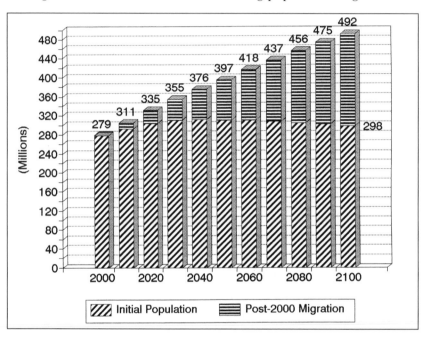

Fig. 9–2. U.S. Population, 2000–2100
(Bouvier/Grant projections, 1994)

The Census Bureau has adopted a rather similar projection. Its middle projection for 2050 is 394 million, a rise of 50%. (They raised that estimate by 100 million after Congress passed the Immigration Act of 1990.) Their high projection reaches 518 million, a doubling, by 2050.

If fertility and immigration stay where they are (figure 9–3 below), the figure goes up to 440 million by 2050 and passes a half-billion long before the end of the century.[154] Growth will be even faster if fertility rises. Most migration is coming from societies more fertile than ours. Most projections assume that migrants' fertility will decline to the present general average but there is just as good reason to assume that they will raise that average as they and their descendants come to constitute a larger fraction of the population (see p. 209 on "shifting shares"). That is what is happening in California, the largest receiving state.

These are projections, not predictions, but they raise a note of caution: *If you are headed somewhere, you may get there.*

Whichever projection you choose, it begins to put us in a league with China and India, and we would be far more destructive because of our consumption levels.

MIGRATION

U.S. data on migration are remarkably bad. Illegal immigration is by its nature uncounted. The Census Bureau makes heroic efforts to include illegal immigrants in the census, but people who do not want to be seen will avoid both the census and the spot surveys that are used to "validate" the original count. The Immigration and Naturalization Service (INS) stopped trying to count emigrants in 1957, and death records are not matched against immigration data, so residual totals are not available. There are literally hundreds of millions of border crossings by land each year. Most of them are not individually tabulated, and there is no way to sort out how many are permanent entrants or departures. There is no system of identification for citizens and legal residents, and thus no effective way to identify who is here illegally. The INS does not count "immigration" by the number of arrivals but rather by the number whose status is "regularized." In other words, the illegal immigrant or overstaying visitor is not counted until his/her status is legalized by such processes as an amnesty, marriage or grant of asylum.

For what it's worth, the Census Bureau figure for annual illegal immigrants (delicately referred to as "undocumented aliens") has been 200,000–225,000 for a decade, and the Census Bureau statisticians recognize that that is almost certainly an understatement. The INS has variously cited figures from 300,000 to 500,000, and its Border Patrol regularly suggests a much higher figure. Airline manifests covering arrivals and departures by air are of uncertain accuracy, but they suggest that illegal immigration by air alone (people overstaying nonimmigrant visas) is several times as high as the official estimates of all illegal immigration.[155]

The Center for Immigration Studies undertakes to estimate the actual annual totals, using Census Bureau and INS data. Here are its results:

FY1992 – 1,183,000

FY1993 – 1,272,000

FY1994 – 1,268,000

FY1995 – 1,206,000

(Fiscal years end on September 30 rather than December 31.)[156] Those figures include 300,000 illegal immigrants annually. As I have suggested, that may far understate the real movement. The INS anticipates a sharp rise in immigration in FY1996.

Legal immigration is probably larger than illegal immigration. Any effort to deal with population growth must address both. Keep this in mind because, as we shall see, there is a widespread perception that only illegal immigration needs "fixing."

Migration is theoretically a less important variable than fertility in determining our demographic future. That is to say, a 10% change in immigration would change U.S. demographic projections less than a 10% change in fertility. But figure 9-2 suggests that migration is presently the more explosive variable and necessarily the central focus of any effort to deal with population growth.

ETHNICITY, RACE AND AFFIRMATIVE ACTION

A graph will perhaps best dramatize one of the major demographic changes underway. (This projection takes us to 440 million because I used the "constant fertility and mortality projection" cited above.) Here is how the population breaks down ethnically.

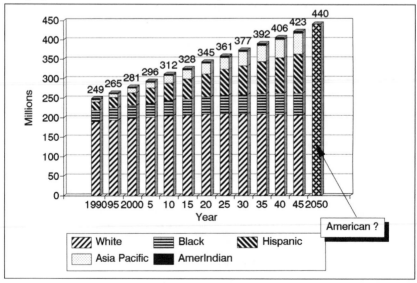

Fig. 9-3. U.S. Population Projections, Constant Fertility, Mortality, Migration
(Grant, 1994, from Census Data)

The "majority," as it is now defined, consists of White males. All others, including White females, are treated as "minorities." In 1990, White males were 38% of the population. By 2045, they will be down to 24%. More and more, we are becoming a nation of minorities—all but one of them requiring special consideration under affirmative action programs. Affirmative action, as we have tried to practice it, is on the way to becoming a victim of demography, as we face the absurdity of trying to give special consideration to almost everybody.

■ ■ ■

Perhaps the time has come to outgrow the concept of classifying ourselves by our ostensible race (a thoroughly unscientific division in any case) or ethnic group. The 2050 column is perhaps a bit whimsical, but it is intended to reflect a hope that racial and ethnic lines will have softened by then into a sense of being American. I believe that we have a choice before us: either we blur the edges of "race" and "ethnicity" or we descend into a warring cockpit of competing minorities. We all have an interest in interracial marriages; they help to make sure that people do not identify themselves as members of one of those smaller tribes. Why is this important? I'll try to answer that question in chapter 15. We must learn to work together as a society.

OUR UNCERTAIN FUTURE

There is of course a danger in taking any specific projection too seriously, but the future is likely to lie somewhere in the neighborhood of those projections. The past is not necessarily a guide to the future, but it is a pretty good bet in the absence of any signs of change in demographic behavior or of a conscious and effective population policy—and we shall see later that there is immense resistance to addressing the population issue.

The developed world and the poorer countries face two very different demographic futures (chapter 7). There is no certainty which way the United States will go—like Europe and Japan, or like Bangladesh and India. We are headed, right now, in that second direction, and liberal U.S. immigration policies such as the Immigration Act of 1990 are pushing us that way.

We do not need to go that way. In chapter 13, I will sketch out the dramatic consequences of a relatively simple pair of policies: "stop at two children," and bring immigration to a level that prevailed for much of this century.

The nation presently has no policy on U.S. population. Immigration policy is dictated by special interest groups—businesses, commercial agriculture, self-proclaimed representatives of ethnic groups and the like—without concern for the national and long-term impact of the migration. Fertility policy is an incidental by-product of other policies. Tax, welfare, health or housing policies may make it more or less attractive for young people to have children and thereby change fertility, but those policies are usually framed in terms of *needs* without considering *consequences*. Do we encourage people to have fewer children or more? We should start looking at the consequences and stop creating our future by accident.

CHAPTER 10

Living with the Land

BECAUSE THE UNITED STATES was sparsely populated well into this century, we have had more space than most countries to absorb the damage we have done. We have not been forced to address our population issues, but we may be more vulnerable than we think. Population growth interacts with the pursuit of national policies in unexpected ways.

The United States is a principal source of most of the environmental problems discussed in Part I simply because of our size and level of economic activity. I will not review all those problems nor try to discuss all the environmental issues in the United States. There are good summaries available (see Note on Sources). Let me explore just three of the principal issues, agriculture, water and energy, and then go on to discuss the popular concept of "sustainability."

AGRICULTURE

The United States may shortly be the Persian Gulf of world food supplies. We supply nearly half the grain exports in world trade. It is an enviable position—for a time—but there are threats to our ability to sustain exports.

Atlas the Uncertain

I have already described the growing third world dependence on imported food, much of it from the United States. U.S. grain yields have been stagnant for a decade. I have argued (chapter 2) that they may not rise very much in the foreseeable future. The United States' productive capacity is not infinite. In the past decade, our self-sufficiency index has fluctuated in the 1.20 to 1.40 range for cereals and for total foodstuffs. That

.20 to .40 represents the margin available for exports. What are the prospects for keeping it up?

Land

With only 1.4 Americans per hectare of arable land, the United States is among the most fortunate countries on Earth, but that ratio is getting less favorable as population grows. By 2050 there will be more than two persons per hectare—still very favorable, but on a continuing decline. U.S. agriculture a century ago moved off poorer lands to exploit the richer prairie soils. We could go back, at the cost of lower yields, and use more fertilizer to make up for the poorer soil. This of course would intensify our air and water pollution problems, which would become more like those of Europe.

The fossil fuel era gave our eastern woodlands a temporary boost. Tractors were more efficient than horses, so pastures reverted to woodlands. That shift and the westward movement of U.S. agriculture explain most of the return of hardwood stands in the northeastern United States and the pine plantations of the southeast that now are a major source of wood products. That respite is finished. The hardwood forests in particular are now subject to urbanization. Putting old soils back to the plow would take back more of those forests and eventually diminish the supply of lumber and of cellulose—just when we need more of both to provide for a rising population and to replace petroleum as a chemical feedstock (see chapter 2).

Erosion

On average, allowing for replenishment, U.S. farmland loses about 12 tons of topsoil per hectare each year (p. 18), and that is better than it was before the most erodible soils were taken out of cultivation by the Conservation Reserve Program (CRP) enacted in 1985. The improvement is welcome, but the loss remains ominous. As this is written, the CRP is under attack both because of budgetary constraints and because rising cereal prices have led major commercial users to demand that we encourage more production. In response, the USDA decided in early 1996 to allow farmers to put some of the CRP land back in production.[157] If world grain prices continue to rise, much of that land will be put back to the plow. We learned from the 1930s dust bowl that you cannot safely make farmland out of rangeland. That lesson is being forgotten. High world demand for U.S. grain will tempt us to repeat our mistake.

We should ask ourselves, Is our object to see how long we can run the system down? or to put our agriculture on a sustainable basis? We still have the option, but we are not geared to think in terms of long-term sustainability. The Department of Agriculture's Universal Soil Loss Equation (USLE) measures a quantity identified as "T." Even many agronomists think "T" represents the level of soil loss that is being replaced by natural processes. It does not. It is an estimate of what a specialist considers an acceptable rate of loss. If the soil is deep, "T" goes up. In other words, it is acceptable to lose the soil, if you do it slowly. That is hardly a prescription for sustainability.

Sustainable Farming

Sustainable farming may preserve the land and diminish the off-farm environmental damage, but it will not be introduced unless it pays for the farmer to do it. There is not yet a clear general answer as to whether it pays; it depends on the place and the farming practices, but some studies show that good conservation practices can benefit the farmer.[158] Even in the best case, however, conservation practices pay because farming costs are lower, not because output is higher.

With better conservation, total output may drop even as the farmers, the land and the environment fare better.

To keep our agriculture sustainable for the long term, we will face a trade-off in loss of immediate production.

Our federal farm support program has been under attack, and justifiably so. It has rewarded the worst kind of agriculture. It was based on gross output and it encouraged monocultures to maximize the farmer's "base" for subsidies. The "Freedom to Farm Act" of 1996 changed the system but did not improve it. Most acreage controls were scrapped and a subsidy was promised for seven years. After that, the supports are theoretically supposed to end, but in the fine print the act simply provides for its own expiration, and the bad old Agricultural Act of 1949 remains in force.

We must change our agricultural rewards system so that at least it does not reward bad farming. Better yet, change the entire system. Penalize monoculture; promote crop rotation, windbreaks, better conservation, and "integrated pest management" (i.e., nonchemical approaches to pest control).

■ ■ ■

To summarize: the United States is not in imminent danger of running out of food for itself, but it has not been able to put its agriculture on a sustainable basis—one that would work for the indefinite future.

Worldwide demand is spiraling upward, and it is tempting to ride the boom, but we should be giving U.S. agriculture a bit of room to reform. Central to such a policy would be a change in U.S. subsidy programs to reward conservation rather than exploitation.

At some point, if our population keeps rising, the United States will have to lose its role as the residual granary to the world (p. 36), and eventually we may not even be able to feed ourselves.

To those who say "we believe we can find ways to meet rising needs," I would respond:

- "Don't count your chickens yet," and
- "Would you be more confident if population were not growing?"

I think the answer to that second question is inescapable: yes.

WATER

Water shortage is a present and mounting problem in the Southwest, including most of California, and in the plains states as aquifers are depleted. I touched on the U.S. situation and our overuse of water in chapter 2. I include it here simply to make two points.

First, we are on a treadmill as we develop increasingly complex ways to conserve water. In most Western states, there are already detailed and expensive procedures to assign water rights, adjudicate differences and enforce the rules. Beyond that, states and cities create and staff water conservation programs. The U.S. Bureau of Reclamation funds conservation projects. Meetings are held and strategies proposed. A 1995 meeting of the American Water Works Association announced technical sessions on "water reuse, interior plumbing and retrofit, leak detection and water accountability, water savings analysis, state conservation policies, alternative financing and rate impacts, and a . . . software package designed to help water utilities evaluate conservation strategies."[159] Temporizing in the face of a basic imbalance is a costly and frustrating process.

Second, since nobody is addressing population, such measures are necessary. They can also be expensive and onerous, and they promote tensions as governments impose new restrictions and interfere in people's lives to allocate the diminishing per capita supplies. Add to that the growth of regulation and policing to enforce environmental rules and public order, and one can understand the widespread current rebellion against government "interference" (chapter 15). Nobody asks, Do we need to do it this way? Shouldn't we take a look at the number of people the resource can comfortably support instead of creating these bureaucracies to manage scarcity?

ENERGY

The energy transition is not simply happening somewhere else. The United States is the major consumer of fossil fuels. We tend to think of our country as less vulnerable than others because it has traditionally supplied most of its own energy. We forget that because we got into the energy binge early, we are the one major consumer that has already used up more than half our original endowment of oil and gas (look again at figures 3-1 and 3-2). Our vulnerability is rising to match that of Europe and Japan, and our government is digging us into a hole because it has no mechanism—no foresight process—to plan energy policy and weigh the impacts of our population growth on it.

Oil

Let us apply the energy calculations of chapter 3 to the United States. The 1994 estimates place known petroleum reserves at 51 billion barrels and unproven resources in the range of 29 to 62 billion barrels. At the anticipated 1% annual projected increase in demand (EIA projections for 1990–2010),

*we have the equivalent of 10 to 17 years' consumption still
in the ground.*

New estimates will probably place the total resource close to the upper edge of these figures as a result of changes in definitions, recent research and improved extraction technologies, but the general conclusion is not likely to change.[160]

Gas

The central estimate for gas resources, modified to reflect increased projected consumption, works out to the equivalent of about 36 years' consumption. As with oil, imports will extend the time, but probably not as much. Gas is not easily moved across oceans, and Canada is already showing resistance to unlimited exports.

Coal

Coal, figured the same way, could last for 156 years, but not if it alone is called upon to replace oil and gas, and not without expensive pollution controls. The country has responded to the oil crises of the '70s by using 60% more coal, the dirtiest of fossil fuels, than in 1970—without enforcing rules that would make coal a truly clean fuel but a costly one. This makes it much harder to control our carbon dioxide emissions and avoid global warming.

Overall

We currently depend on imports for 22% of our national energy needs, compared with 6% in 1970. For petroleum, we now rely on imports for more than half our consumption. That figure will keep rising. The big oil companies have pretty much given up on the United States other than Alaska. They are shifting their exploration energies elsewhere.

Does the U.S. Government look with equanimity on our increasing dependence? It moved very quickly when Iraq threatened to destabilize the Arab peninsula. I have pointed out the dangers of a recurrence of instability (p. 49). We could come under immense pressure to mount a "peacekeeping force"—very possibly against Arab opposition—to try to keep the wells pumping and the pipelines open. Not a pleasant prospect.

We cannot safely rely on a functioning world system to provide essential necessities such as energy. Nor can we rely solely on our own conventional fuel resources. What is the answer? Stop the rising demand, and make renewables competitive.

We need to diversify our own sources of energy to keep net imports within tolerable limits. If we diversify toward benign renewable energy, it will prevent serious environmental problems. Diversification should be near the head of our national priorities. It will be an expensive change, involving high capital costs. Insofar as we turn to biomass, it will compete with food for land and eat into our export capability. We must aim for the "new renewables" such as wind and direct solar energy.

Meanwhile, Senator Murkowski (R–Alaska), Chairman of the Senate Energy Committee, is pressing to open the Arctic National Wildlife Refuge to exploration. With luck, we may gain a year or two of consumption but at the expense of the kind of mess typified by the *Exxon Valdez* disaster.

What do we do?

The United States, more than any nation, needs a *pollution tax* or fossil fuel tax. If we had the discipline that OPEC lacks, we would raise taxes on fossil and probably nuclear energy and capture the windfall for ourselves. A systematically rising tax on fossil fuels would achieve several objectives:

- raise the cost of such fuels so as to make conservation more attractive and to make alternative renewable fuels more competitive—without having to subsidize them;
- cause automobile users and industry to absorb a larger fraction of the true costs of their activities and to adjust their activities so as to minimize fossil fuel use;

- thereby reduce the sources of air pollution, acid precipitation and global climate change;
- save foreign exchange; and
- generate revenues to balance the national budget.

Such a policy would prepare the nation for a gradual entry into the forthcoming energy transition. The tax would encourage an evolution in our agriculture to adapt to a less energy–intensive future. It would promote a sustainable ecology and a revised human landscape to accord with such a future. It would help generate capital for the transition. Having a "proactive" policy (in the new jargon) is better than simply waiting for the next crisis and then belatedly scrambling to make the adjustments.

A fossil fuel tax (or "carbon tax" or "pollution tax") would be an equitable way to balance the budget. Granted, such a tax would probably have a disproportionate effect on the poor, because they spend a larger share of their disposable income on transportation and household heating than do the rich. However, properly managed, the tax would embody some compensatory tax relief for the poor without subsidizing their fuel consumption.

Such a tax is one of the few proposals on which environmentalists (including the IPCC), academic economists and budget specialists agree. The problem is selling it to the American public. As this book goes to press in an election year, even the existing federal energy tax is under attack, and the Republicans propose to roll it back.

If we were wise, we would accompany any pollution tax with a population policy. I have mentioned the pledge at UNCED to hold down our carbon dioxide emissions. The other industrial countries, with populations more or less stable, need simply to find a few more efficiencies. The United States is fighting its way uphill. Each decade, our population growth imposes a 10% growth on energy demand. The task of limiting those emissions keeps getting harder.

There is considerable agitation (though not within Congress) for better conservation practices to hold down demand. We are trying to become more efficient in our use of energy. We have made progress. We use only 60% as much energy per dollar of GNP (in constant dollars) as we did in 1970. Part of the gain has been because of the growth in services, which don't generally use much energy. Very good, but you can carry that process only so far. You cannot eat services or wear them or live in them. Our energy use has risen 27% since 1970—right in step with our population growth—despite the increased efficiency. We don't need that hurdle.

Japan is often cited as the model of efficiency. It is, but it is not so far ahead of the United States as some comparisons suggest. On their tight little archipelago, living in tiny houses and flats, with a population dense enough to support public transport, distances too short for aircraft and not much room to drive or park automobiles, Japanese use just over half as much energy as Americans do (adjusting for the difference in GNP per capita).[161] We can move in that direction—in theory—but it would take some fundamental changes in living habits, and it would take time. Having gotten people out of their energy-intensive cars and into smaller houses, the question would arise, What do we do next?

The Japanese have answered that question for themselves. They are approaching zero population growth—somewhat to the consternation of their traditionally minded government. We too should be looking at population growth as part of the energy issue, yet nothing is being said about the role of population growth in increasing demand or how that relates to the present debate over immigration.

A deliberate effort to reverse the growth of the demand side—
population—would be the cheapest energy policy.

Such an effort would lead to rising efficiencies, as the pressure on our resources diminishes, rather than raising costs as we try to squeeze every ounce of energy out of nature. It would help to achieve the goals of a pollution tax (above), and it would avoid the prospect that when another energy crisis comes, government regulation will be "on our back" to save a desperately needed resource.

SUSTAINABILITY: THE RIGOROUS GOAL

The word "sustainability" has appeared several times in this book. It means the maintenance of environmental and resource systems so that their ability to support future generations is not diminished. It is a popular concept right now but still a somewhat fuzzy one. There is, for instance, no real agreement as to whether it refers to humans only, to the preservation of a healthy biosphere or to all species—which would be quite an undertaking, since species have been appearing and disappearing since the beginning of life.

There is a danger that enthusiasm for the concept of "sustainability" can lead to absurd rigidity. Zero soil loss is not really compatible with any agriculture; even if it is replaced, some of that soil descends into streams, accelerating siltation and change. Change, and perhaps degradation, can occur without human involvement; the Mississippi delta was being formed out of prairie topsoil long before humans arrived. Aluminum constitutes 8% of the Earth's crust. One can contemplate running down the resource with equanimity. The argument for recycling, aside from aesthetics, is the high energy cost of extracting the aluminum from the ore.

Sustainability does not admit of much compromise, and we are inclined to fudge it, as with the USLE and the critical "T" (p. 165). A hydrologist remarked that there is "no agreed definition of what 'sustainability' means with regard to a water table." Yes, there is. If a water table is sinking over time, it is not sustainable. Then the question is, At what rate do we compromise with sustainability? Should we stop mining the water when we reach a rate that would allow 10 years' use of the aquifer? 100 years? 1000? 10,000? One may be sure that such temporizations will be made; in the practical world the only hope is to resist the more damaging ones. Environmentalists may have to be content with pushing the reckoning off into the future, but compromise should not be mistaken for sustainability.

Despite its weaknesses, the idea of sustainability is central to any effort to learn to think beyond the moment. Sustainability may demand zero growth. When one is used to growth, it is very hard to think in terms of zero. Experts have told the city of Santa Fe that it is already using all the water it has and that in a dry period there won't be enough. I was talking about the problem with a friend, an intelligent and active participant in community affairs. I suggested that the city may need to limit new water connections, and his reaction was "good idea; maybe they should be limited to 3% per year." In our society, 3% growth looks small, but 3% per year is a quadrupling in two generations, a 20-fold increase in a century. Santa Fe will be very lucky indeed to find much more water. It is on a collision course. Before very long, it must figure out how to achieve zero growth. Growth itself is not sustainable.

It is even harder to address problems when we are already past sustainability, as the nation is with its topsoil and groundwater supplies, with nuclear and toxic wastes and with its use of automobiles and fossil fuels. Then the question becomes, How do we move back when people need jobs?

Perhaps the toughest issues of all are biodiversity and the survival of Earth's ecosystem. Most biologists would probably tell us that we are already over the limits. Habitat extinction means species extinction (chapter 6). The destruction of wetlands is a classic case, and the country right now is moving back from the preservation measures it had adopted in the past two decades—which themselves were proving insufficient to prevent a continuing loss.

■ ■ ■

Is there a way out? The conventional answer is a shrug of the shoulders, but there is a way. Look at the demand side: population. Bringing population growth under control is a matter of will. Given that will, it can be done (chapter 13), and the pursuit of sustainability can become a practical possibility.

CHAPTER 11

The Collapsing Society

OUR SOCIAL SYSTEM is coming apart in complex and interrelated ways that our decision processes are simply not prepared to deal with. Decaying, impoverished cities. The decline of wages. Unemployment. Job uncertainty even among groups that had felt themselves secure. The breakdown of law. The rise of drug use and violent crime among Black kids. The wanton destructiveness by wasted youth. The weakening of the family and the destruction of traditional anchoring values. In the cities, the young feel they have been excluded from the system and, like the children in *The Lord of the Flies*, they create their own violent and malevolent parody. In the western United States, we watch a conflict of value systems so intense that Forest Service vehicles and offices are bombed in Nevada, and a sheriff runs a bulldozer at rangers in a national forest. Terrorists, foreign and domestic, bomb the World Trade Center in New York and the Federal Office Building in Oklahoma City. In Arizona, self-proclaimed vigilantes derail a passenger train full of innocent people.

Things have gotten worse for the poor, and they grow alienated. The middle class has lost its sense of security. We grow aware of our rising energy dependence on distant and unstable regions. The world suddenly seems a lot less secure place.

We go deeper into debt to the rest of the world and (in the form of the Treasury bonds) to our own rich. Our efforts to manage crime and drugs and our policies to produce full employment evaporate in smoke. We are in a whirlpool, and those elements of decomposition are synergistic.

DISINTEGRATION OF THE CITIES

The decay is exemplified in some frightening statistics about deferred maintenance. A Department of Transportation estimate puts the cost of repairing existing bridges at $50 billion, and the cost of restoring and maintaining existing highways at their 1983 condition through 2000 at $315 billion. The Associated General Contractors of America estimate that it would cost $3.3 *trillion* over the next 19 years to repair all of the public infrastructure for the existing population.162 That is nearly half our annual GNP.

The Bypassed Ghetto

Change has moved too fast. Particularly for many Blacks, the move from the rural South to northern cities took place in one generation, and within one more generation the jobs that supported those cities began to disappear.

Our most serious unemployment problem is invisible in the unemployment statistics. We console ourselves with the official statistic: the unemployment rate is less than 6%. Nonsense. That figure helps us ignore the scale of the problem. It counts only active job applicants, not "discouraged workers" no longer beating the streets for jobs. For a more realistic look at what is happening, look at the proportion of those of working age who do not have jobs. Even worse, look at the young. People who

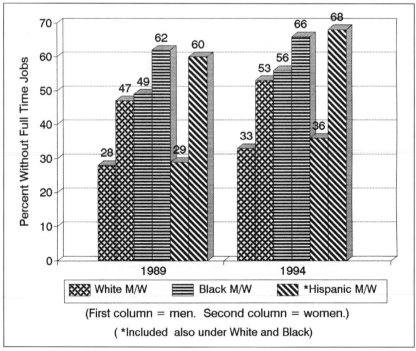

Fig. 11-1. Young and Idle. 16–24 Year Olds Not in School, Jobless
(Bureau of Labor Statistics: Employment and Earnings)

cannot find a job by the time they are 24 are probably going to wind up defeated, resentful, alienated and perhaps violent.

The graph is about the young people who are not in school, not in the military and not already in prison. The proportion who do not have regular jobs ranges from 33% for young White men to 68% for young Hispanic women. Granted, some small proportion, particularly of the women, may be homemakers or voluntarily living with their parents and not working, but by and large, these people have been excluded from the system or have simply not tried to join it. They are getting by on crime and whatever scraps they can earn in the "informal economy." As the comparison with 1989 shows, the problem is not just bad; it is rapidly getting worse. Don't just think of the social costs. Think of the wrecked lives.

Idleness and alienation produce vandalism and crime. There is a public perception of rising violent crime. The statistics are not consistent, but they do not suggest so much an overall rise as a highly concentrated pattern. The crime rate is about four times as high in metropolitan areas as in rural areas; in 1993 there were a remarkable three or more violent crimes for every 100 people in Atlanta, Miami, St. Louis, Newark, Tampa and Baton Rouge (in descending order) and almost as many in Baltimore, Washington, Chicago, Detroit and Oakland. Both perpetrators and victims are most likely to be young Black males, with more than half of all homicide victims Black—34% of all murder victims are young Black males between 15–34 years old, who constitute only 2% of the total population.[163] These figures are perhaps the coldest measure of those wrecked lives.

The Poor Get Poorer

It is not simply the unemployed who are being marginalized. Productivity is rising, while real hourly wages decline. The benefits are going to management and capital, not workers. The Labor Department reported that productivity in durable goods in 1993 rose 7.5%, but hourly compensation dropped 0.2%. A President's Commission on the Future of Worker–Management Relations bemoaned the bifurcation of the work force, creating "an underclass of low-paid labor."[164]

Management can get away with it only because we live at an intersection of trends that has put people and jobs far out of balance. I will present the data later in the chapter.

This is an explosive mixture. It has been observed of the French Revolution that revolutions tend to occur, not in situations of unrelieved poverty, but when things have been better and begin to decline. Right

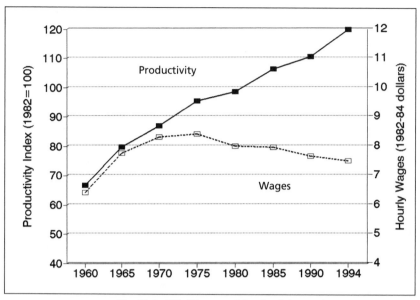

Fig. 11-2. U.S. Wages and Productivity
(*From U.S. Statistical Abstracts*)

now, people are living with outmoded expectations in an era of declining wages.

The Rich Get Richer

A recent article in the *Wall Street Journal*—not usually considered a radical rag—pointed out that in the past two decades the real earnings of the bottom fifth of Americans have dropped 24% while those of the top fifth have risen 10%. Twenty years ago, a company CEO typically earned 35 times as much as the average worker; the ratio has risen to 150, the highest in the industrial world.[165] A multilateral study showed that the income gap in the United States between the richest 10% of the population and the poorest (including government benefits) was $55,000, a ratio of 6:1. Our rich are the richest; our poor are the poorest in any modern economy except for Ireland and Israel.[166] The proportion of our country's wealth held by the rich has been growing. In this country, the top 1% of families own over 40% of the nation's wealth. In England, which is generally perceived as class-ridden, the figure is just 18%.[167] There is a feeding frenzy among our rich.

For those who subscribe to a minimalist view of government, for whom government exists simply to provide for the common defense, maintain order and run the courts, this may be fine. Their satisfaction may be short-lived if the price is social turmoil.

We are moving from a democracy to a plutocracy. Political candidates must solicit huge bankrolls to compete on television. Population growth helps to prostitute the process. Since 1912, when Congress fixed the size of the House at 435 members, the population of the average congressional district has nearly trebled to over 600,000. Senators, who were not even directly elected until 1913, have an horrendous campaign financing problem. Pursuing the presidency demands astronomical financing. Voters can be reached only on television—an unforeseen consequence of population growth and technological change. Politicians respond not to *vox populi* but to potential contributors, trying meanwhile to obscure their subservience by parliamentary maneuver if the public is sufficiently aroused on an issue. (Even plutocracies can be prudent.) It is very hard to suspend disbelief and to trust that somehow, in this one relationship, bribery does not work. A corrupted government and a cynical electorate undermine our whole system of governance.

Lest I myself sound too cynical, I promise you an example in chapter 12 and a detailed account in chapter 17 of the flouting by successive Congresses and administrations of the clear popular will concerning immigration. Congress was responding not to the people but to money.

Are the rich getting away with it? To quote from Rudyard Kipling's poem about the browbeaten and underpaid British soldier: "you bet that Tommy sees!"

The Middle Class Gets Restless

Pressed by the stagnation of their own living standards, harassed by the inability of the cities to provide minimal services and above all frightened by the mood of the poor, the more prosperous of all races and persuasions take flight to the suburbs. Since they are the taxpayers, the urban tax base declines. Urban services further decline, and employment opportunities dwindle as the prosperous take their needs for services with them out of the city.

For the first time since the Great Depression, professionals and skilled technicians find their jobs in jeopardy because of corporate downsizing, the export of jobs and corporations' importation of lower priced third world professionals to replace them. I will run some of the numbers later.

This isn't the end of it. With nerves frayed by job insecurity, the middle class' envy of the more fortunate is bound to grow. Our society is being pulled apart by the centrifugal forces of division and alienation. *The*

Washington Post recently ran an in-depth report on middle-class Whites dissatisfied with the political system who are creating networks to explore radical approaches. It ran another article about middle-class Blacks who are turning conservative but, more important, are turning inward toward the Black community and warning it against expecting help from others.[168] Both attitudes reflect a deep sense that our system is out of whack and that our traditional governmental approaches cannot cope.

Since dissatisfaction with the current state of affairs is no assurance that the dissatisfied know a better way, the prognosis for a happy society is not hopeful.

THE SOURCES OF UNEMPLOYMENT

Unemployment is linked to a synergistic set of changes that drive unemployment upwards and wages down. Those changes are —

- a third world population explosion that has driven their working age population up by nearly 1.7 billion since 1950 and will add another 2 plus billion by 2025. It is causing a new "Age of Migrations" from farm to city, across national boundaries and spilling over into the United States and Europe (chapter 8).

- a technological transformation of work (chapter 5) that demands fewer and fewer people just as there are more people competing for work.

- a restructuring of world commerce that very nearly assures that wages in the high-wage countries will decline and unemployment will increase (chapters 5 and 8).

- a social transformation that has expanded the work force as women seek employment and, with it, a sense of dignity and independence.

- a diminishing resource base for the extractive industries.

Let us take a closer look at each of those forces.

The Migration of the Hungry

I have described the extraordinary population pressures in the poor countries (chapter 7 and figure 8-1). Their problem becomes our problem. I do not mean to suggest that those billions of people will all suddenly decamp for the United States, but consider this hypothetical case: if one-tenth of them should move to the United States in a generation, the migration would absorb less than one-third of their natural increase, but it would more than double our population. That presumably would make the United States about as attractive to job seekers as India, thus discouraging the rest. Hypothetical, but Europe sent a much larger proportion of its people to the New World in the nineteenth century.

Less hypothetically, the United States' working-age population (16–64 years old) will rise by 42% to 59% by 2050, depending upon which of the projections in chapter 9 one uses, and most of that results from immigration.

It is not just the hungry who will come. There are plenty of educated and underemployed technicians in India, for example, who can earn a lot more here. U.S. business leaders like that idea. For businesses that need to obtain employees quickly, there are go-betweens who specialize in providing Indian technicians such as computer programmers, on a contract basis. The business does not need to inquire just how they got here. They are likely to be on a tourist visa. For the more fastidious businessmen, there are two sections of our immigration law that enable businesses to import such labor—or they can hire persons who came under other categories (see figure 12-1). Supposedly, they must import only specialists in short supply in the United States and pay them the going wage, but there are many ways to circumvent the rules.

The Immigration Act of 1990 raised the quotas for work-related visas to 140,000 to meet businesses' demand. Business leaders argue that they must have that labor to stay competitive. In part, that argument is circular. If everybody is doing it, you must do it, too, to stay competitive. At another level, however, the business leaders have a point, and it goes back to the conundrum in chapter 8 about preserving wage differentials in a world of poverty. If they don't import the technicians, the jobs will go overseas. Either way, U.S. technicians lose jobs. We cannot escape the free trade conundrum.

The nation must make a very tough decision: Do we want to try to preserve American jobs and wages if that means a tougher view of immigration, controls on trade and the willingness to be a relatively prosperous island in a world of intensifying poverty? It is not an easy choice, but it must be made and acted on.

Technological Change

The problem of technological unemployment has come home to the United States with a bang: computerization, new managerial techniques, automation, the movement of some industry overseas and now the efforts of corporate America to control its labor costs. Let me add some numbers to my earlier statements (chapter 5) about the impact of technology on employment.

The Census Bureau's *1992 Census of Manufacturers* shows what technology is doing. Manufacturers' shipments rose by 21% from 1987 (not adjusted for inflation) while capital expenditures rose 32%, but employment fell 4%. These statistics help to explain figure 11-2. Invest in capital, not in workers, cream off the increased productivity, and let fewer and fewer people do more and more of the work, while the others are off your payroll. In this environment, even high-salaried people come to need help.

Firms are downsizing to limit their vulnerability to business-cycle fluctuations and technological redundancies. They are contracting for temporary help when necessary. A private survey indicates that corporate downsizing is resulting in the loss of more than a half-million jobs every year, and these losses occur among those who had been the best paid and the most secure.[169] The second biggest employer in the United States, after General Motors, is Manpower, Inc., a firm that supplies temporary workers, or "temps," to businesses that want to escape their responsibilities as employers. Business' own organization, the Conference Board, ran a survey and learned that 21% of the companies responding now use temps for 10% or more of their staffing, almost twice as many as in 1990, and more than a third expect to pass 10% by 2000.[170]

On top of that, both political parties propose substantial reductions in federal employment, which has been static for decades.[171]

A choice lies ahead: massive income transfers from those in the system to those outside it, or chaos and domestic strife. Very clearly, the country is in no mood to entertain the first idea or to contemplate the alternative.

Free Trade and the Multinationals

I have touched upon the connection between trade policies and U.S. jobs and wages. Job competition and the labor market have gone international (chapter 8).

Part of the problem has arisen almost unnoticed: the MNCs' role in fostering that competition. They have fundamentally altered the world trading system, yet U.S. conventional wisdom is unchanged, and we are trying to play by the old rules.

There is much anecdotal evidence of job flight. One *New York Times* article followed the effects on employees as Texas Instruments began designing sophisticated computer chips in India for one-third to one-fourth the salaries paid in America; Motorola Corporation set up equipment design centers in China, India, Singapore, Hong Kong, Taiwan, Australia and South America; and CSX Corporation closed a programming operation in New Jersey to open new ones in India and the Philippines—and even asked the fired employees to stay on to train their replacements. American International Group did the same. A Columbia University professor remarked, "You're definitely seeing an enormous integration of the markets, and therefore a drag on the real wages here of the semi-skilled (and) . . . skilled."[172]

I repeat the conundrum I raised in chapter 8: How do we maintain living standards in a world with free movement of capital, technology, management, marketing systems and goods? The multinationals can produce where the labor is cheap and docile and then sell the goods back here. Eventually, we won't be able to afford them, but then the MNCs will find something else to do.

Any proposal to address the problems of joblessness and alienation is dead on arrival if we do not take another look at the American free trade shibboleth. We will look at the issue in the next chapter.

The Role of Women

Throughout the world—including the United States—the nonagricultural labor force has grown as more women seek paid work. One can hardly begrudge them the opportunity, since with it come self-respect and a sense of independence. And they need the money.

In the United States, the proportion of women over 15 years old in the labor force has risen from 39% to 59% since 1964, but it is still below the rate for men (75%). Some of those others may seek jobs, particularly as declining real wages cause them to seek work to maintain family income.

The Diminishing Resource Base

Economists used to list "land" (i.e., resources) along with capital and labor as the three factors of production. "Land" was forgotten as economists adopted the convenient myth of infinite substitutability of resources.

"Land" is still important to varying degrees in different sectors, even though the growth of the service sector makes resource-intensive sectors a diminishing part of the GNP. We are watching a particularly poignant confrontation in the Northwest, where loggers, spurred on by timber interests, are in a fighting mood to keep on harvesting old growth forests. It is a triumph of myopia over foresight. (One pungent cartoon showed a logger and a supervisor in a bare landscape of stumps. The logger has just cut down the last tree. He turns to the supervisor and says, "Well, that does it." The supervisor says, "You're fired." I would add that the supervisor will go next.)

The loggers have a problem. There isn't much else to do with their skills. But the problem is much larger. Logging, hydroelectric dams and overfishing have caused the collapse of the salmon industry. Northwestern fishermen are without jobs—and mad at the loggers. An effort was made to help save the salmon by releasing more water from reservoirs at salmon migration time. This affects the hydroelectric power that has been a source of cheap electricity and jobs in the Northwest, so industries and cities are incensed at the proposed solution. Thus, a compounded resource problem leads to a triple confrontation and threatens unemployment in multiple sectors.

A policy to hold down or reverse U.S. population growth would have meant less pressure to cut the forests, a less intense demand for electric power, more fish, and sustainable jobs for the loggers, fishermen and industrial employees of the Northwest.

You can push resources only so far. Jobs still depend on the resource base.

■ ■ ■

Are we such fools that we cannot recognize the law of supply and demand when it comes to labor? Or the source of the rising prosperity of the 1950s and 1960s, when the small depression–era cohorts entered the labor market and enjoyed an unparalleled rise in living standards because they were in a sellers' market? There is an historical parallel of interest: in nineteenth–century Europe, in one country after another, wages rose as many workers migrated to America, and then emigration slowed as wages rose in the sending countries and the gap closed.[173] A direct demand and supply equation. The problem is now too large for us to provide that relief valve again (p. 245).

THE FATEFUL INTERSECTION

If one undertook to launch a synergistic set of destructive trends, it would be hard to improve on the set presently operating in the country:

- Those five different forces (above) making access to jobs more difficult.
- A decline of wages after a period of rising wages and high hopes. A parallel and sudden sense of insecurity among the educated elite as big employers "downsize" and "outsource," and they are left with no sense of personal security or attachment to the companies for which they work.

- The invidious comparison as workers watch the rich engaged in their feeding frenzy.

- Color television in almost every home, to make sure the comparison is not missed.

- A world grown complex beyond comprehension, in which very few can understand the functioning of the technologies on which they rely or the economic system in which they live, coupled with a national focus on ethnicity that encourages the dispossessed to lay the blame on race rather than economics.

- A pervasive sense of insecurity resulting from reports of violent crime on television. For the more thoughtful, the sense of a deteriorating and insecure world and of dependence for energy and jobs on distant and unstable regions.

- Disenchantment with our political leadership, resulting from the widespread recognition that our leaders respond to money more than to the hopes of the electorate or any perception of the national interest. The natural response is, Why pay taxes to support those bastards?

The insecurities vent themselves in rage. It will be worse if the job market tightens and if wages decline farther. If we follow our present policies, those things will happen.

■ ■ ■

It would be disingenuous to suggest that a population policy would cure all these problems, but a wiser immigration policy and trade policies that do not penalize American labor would be a start. A sane population policy would help match our people to the jobs and our population to the resource base. We need to restructure the democracy to save it from the power of money and to persuade the owners of the country (the rich) that a bifurcated society is not in their own interest. The bitter dispossessed must get back into a frame of mind where they can be productive members of society. Quite a list.

CHAPTER 12

Population and Policy

WE CANNOT, AS WE HAVE DONE BEFORE, undertake to deal with our problems in isolation. The government is trying to shore up one crumbling structure after another when we need an overarching vision of what the nation must do and what we can afford to do. In this chapter, let me offer some thoughts about what we can do to ameliorate the problems and how our choices are limited and shaped by population change. Because we have not been willing to take on the population issue, we must for a long time live with the results and try to accommodate our policies to the realities of demography.

The solutions may be compromises that we did not expect to have to make. The illusion of omnipotence and unlimited possibilities began to crumble sometime in the '70s, and it will not soon come back.

IMMIGRATION AND JOBS

There is a tremendous inertia in our decision-making processes. The graph below gives one dramatic example. With job growth stagnant and the U.S. working-age population growing by 1.5 million per year, wouldn't a rational society avoid intensifying the competition for those jobs? Yes, but decisions in this country are not made rationally. They are made in different cubbyholes in Congress and the executive branch, and such questions don't get asked.

As the graph shows, our immigration policy introduced many more job applicants into the economy in one year (FY1992) than the 256,000 jobs created by the economy in three years of stagnation.[174] Such disconnected decisions pervade our immigration policy.

More important, they reveal governmental unconcern about unemployment and reflect the priorities of our emerging plutocracy. The Chicago Council on Foreign Relations for years has sponsored periodic Gallup

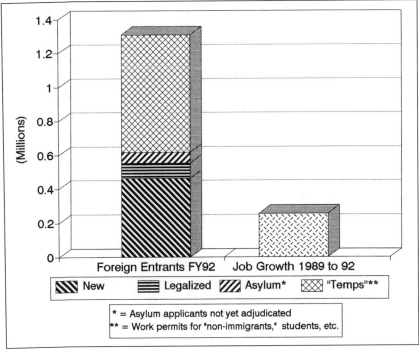

Fig. 12-1. Alien Work Force Entrants vs. Job Growth
FY92 Entrants and 3-Year Job Growth

polls on national attitudes relevant to foreign policy. They are unique, so far as I know, in reporting the results separately for "the public" and "the leaders" (political, governmental, business, media and education leaders "in senior positions with knowledge of international affairs"). The results are fascinating. In 1991, protecting the jobs of American workers was listed as a "very important" foreign policy goal by 65% of the public respondents—the highest of any topic identified—but by only 39% of the leaders. And 60% of the public but only 30% of the leaders believed that foreign policy had a major impact on U.S. unemployment.

By 1995, a remarkable 84% of the public but just 51% of the leaders saw the protection of U.S. jobs as an important foreign policy objective. There were astonishing gaps in the attitudes toward immigration. Immigrants and refugees were seen as "a critical threat" by 74% of the public respondents but only 31% of the leaders. Controlling illegal immigration was seen as a "very important goal" by 73% of the public but only 28% of the leaders.[175]

I think the people may be right.

I will discuss the results in chapter 17: the defeat of periodic efforts in Congress to lower legal immigration and to bring illegal immigration under control, abetted by the equivocation of Democratic and Republican administrations.

What about immigration and the urban drop-outs?

Years ago when I was in Taiwan, faith in capital-intensive solutions was at its peak. If you need a logging road, said our aid program, bring in machinery and build it cheaply so that the costs stay low and the lumber is competitive on the world market. At that time, there were tens of thousands of retired servicemen in Taiwan who had fled the Chinese mainland with Chiang Kai-shek's government. They had to live. One Chinese leader with vision and a sense of responsibility to the retired soldiers refused to mechanize, saying "I'll put the veterans to work with shovels and dynamite, and I'll pay them enough to live."

There is a parallel in modern America. The immigrant is the analog to heavy equipment. Employers are delighted to hire immigrants who know what it is to be hungry and who are satisfied with a wage that may be rock bottom in the United States but is princely in Mexico. If the immigrants are illegal and the employer unscrupulous, he can work them harder by threatening to turn them in. The problem is that we have the unemployed. We cannot simply let them waste away, and in this society they are not wasting away quietly. Employers point out that those people, most of them young, really can't work. They have never learned the discipline. The employers are probably right. It is not just a question of jobs; it involves education, discipline and socialization. But the other elements fail if there are no jobs.

■ ■ ■

The solution is not simply to find some way to motivate those young people. It is difficult to motivate people if there are no jobs to motivate them into. Financier Felix Rohatyn (best known for his role in rescuing New York City from bankruptcy in the '70s) has called us "a nation of outcasts in a nation of wealth" and asked what is the purpose of retraining and education if there are no jobs.[176] A very good point.

If there were more jobs, exhortations to the young to learn to read and write and get to work on time would have more meaning, because there would be jobs at the end of the effort. But, if Wassily Leontief is right and if the present trends continue, there won't be that many more jobs.

If there were fewer immigrants, there would be a better balance between jobs and job seekers. If, through family planning, there were

fewer unplanned and often unwanted babies, there would be fewer disenfranchised youth, and the competition for available jobs would be less
ferocious. With fewer people to educate, funds for education would not
be spread so thin, and the young people would be better equipped to handle the jobs that exist in this sophisticated economy.

FREE TRADE AND AMERICAN JOBS

The government must eventually address the conflict between American trade policy and the American job market. Joblessness is a deadly
condition in the modern world (chapter 5). Labor policy should be the
central determinant of trade policy. It is not now even an add-on.

Somebody once remarked that generals are always fighting the preceding war. Many economists are still fighting the Great Depression. I
came through that era, and I believed what was taught me: to hate the
Smoot–Hawley tariff act, which helped to precipitate the Great Depression,
and to believe that free trade is sacred.

We become furious with Japan because it keeps fudging on the principle—our principle—of open trade. (Of course, we fudge too, when things
get tough. Witness the history of "voluntary restraints" imposed on our
trading partners' exports of textiles, shoes and automobiles, and import
controls on sugar and peanuts. A foolish consistency, said Emerson, is the
hobgoblin of little minds.)

Some economists are now questioning the faith in free trade. Japan's
balance of payments surplus has led some to suggest the possibility of
managed trade. They may be shooting at the wrong target. Japan is certainly a current problem, and its trading practices may force us into some
trade management ourselves, but I have suggested (chapter 8) that there
are other trade threats to follow Japan.

Mr. Clinton, Meet Mr. Clinton

In his 1994 State of the Union Address, the President said, "We can't renew
the country when our businesses eagerly look for new investments and
new customers abroad, but ignore those people right here at home who
would give anything to have their jobs and would gladly buy their products if they had the money to do it." (He was applauded.) Why didn't he
think of that when he was twisting arms to pass the NAFTA treaty, much
of which is devoted to facilitating the movement of U.S. investment capital into Mexico?

My concern is not for the multinationals. They may well profit from
moving production to Mexico, at least in the short run. I am worried about

whether the government of all the people should be investing its effort to facilitate that exodus of capital.

Immigration is a threat to U.S. wages, but the downward pressure on wages will worsen with free trade whether or not the nation reduces immigration, until U.S. wages are so bad they are competitive or until some distant day when third world population growth is under control and wages in those countries have risen to decent levels.

President Clinton's ambivalence also throws a brilliant light on the power of money in our system. Why did he fight to pass NAFTA if he was worried about American jobs? I think I have already answered that question.

Our hand may be forced by rising European and Japanese barriers. If we try to survive as the last bastion of free trade in the industrial world (p. 144), the pressures on our jobs will rise, unemployment will go up, and the nation will not have the means to cushion the shock with welfare.

The Way Out

I propose that we abandon our fixation on free trade, offensive as the proposal may be to many (including perhaps some readers of this book) and as politically unrealistic given the present mood in Washington. I urge that the country establish a new priority and consider first the impact of trade policy on American workers. The world trading system, as presently organized, offers no assurance of protection to our workers, just as the world energy market offers little energy security (chapter 3).

Trade is a particularly tough nut, conceptually. A sharp rise in U.S. tariffs would plunge many exporting countries into turmoil. The problem is that with our free trade shibboleth we have allowed the dependency, particularly of the emerging countries of Asia, to reach such a level (chapter 7) that there is no graceful or easy way to help them out of it. They have prospered from export–oriented economies dependent on the U.S. market and from the United States' ability to supply their food needs indefinitely. We should have been warning them against the dangers of such dependency, just as we should ourselves be cautious about our own energy dependency. If both we and they had thought ahead, these would have been recognized as central issues long before now.

A rise in tariffs would raise prices in the United States and exacerbate unemployment in our exporting industries. Altogether it is a delicate balance, but the present course leads us all to disaster.

I would argue that we can best preserve a rational world trading system if we move gradually—starting now—rather than allowing free trade to drive us to a crisis. Demographic pressures that most economists do not recognize are becoming intolerable. The system needs, not massive tariffs, but enough impediments to the movement of industries and jobs to provide labor with at least a transitional protection against sudden displacement.

Perhaps a solution lies in some sort of agreed "wage–differential tariffs" keyed to the labor content of the product, thus protecting labor-intensive industries—but that would drive industries elsewhere to substitute capital for labor and worsen the world unemployment issue. We need a fundamental reorganization of the ITO, perhaps to substitute some sort of managed trade policy. We should start with a frank appraisal of our own interests but recognize the potential impact of U.S. trade restrictions on the stability of other nations. Any managed trade policy would need periodic modification. (There is a parallel of sorts in Canadian and Australian immigration policies.) The problem with this idea is, Who would manage the trade? Current experience suggests that it would stay in the same hands with the same old priorities. I confess to bafflement.

■　■　■

To reiterate the connection with population: we are in a world with not enough jobs, where working–age populations are growing, but perpetual economic growth to provide jobs is no longer a possible solution. The American demographic behavior that intensifies the domestic competition for jobs—high fertility among the poor and high immigration—is further exacerbated when the competition becomes worldwide. And the phenomenon of growing populations dependent on foreign markets and resources can suddenly become a deadly two–edged sword.

In an ideal world, free trade would be an attractive idea, like free movement. The way to that ideal is not to pretend that we have arrived there but rather to preserve decent labor standards where we can and to help third world countries to reduce their growth. If they succeed, their labor force will eventually stop growing, and perhaps the wage gap that makes free trade presently impossible will close or narrow. U.S. foreign aid policy (chapter 13) could make a contribution to resolving the dilemmas we face in trade policy.

JOBS AND WELFARE

President Clinton in the 1992 election campaign and the House Republican "Contract with America" in 1994 both promised to put welfare recipients to work. The idea is politically popular. Welfare is resented by both the donors and the recipients. The problem is that those five synergistic forces—population growth and immigration, the internation-alization of production, technological change, the entry of women into the work force, and the diminishing resource base per capita—are driving the nation toward higher rather than lower welfare dependency. Any realis-tic welfare proposal must be linked with population, immigration, trade and technological policy. We may be pretty confident, for instance, that the present downward pressure on jobs and wages will continue until we address the labor impacts of free trade. And jobless people need welfare.

Recognizing those connections, let us look at welfare to see how the limits of the possible will determine what we can do. The topic is on and off center stage in Washington, and it is difficult to shoot at a moving tar-get, but I think there are things to be said about welfare that are not presently being said.[177]

In its health care package, the Clinton administration sought to expand the coverage, whereas in theirs the Republicans seek to control the costs. With welfare, both Democrats and Republicans are trying to get people off welfare and cut costs. It sounds easier, but it isn't.

Education as Solution?

The President and his welfare task force proposed to educate people to make them employable and get them off welfare. The United States has sunk to a miserable condition. A recent Department of Education survey indicated that one-quarter of adult Americans could hardly read and could not add. (A quarter of those people, by the way, were immigrants still unfamiliar with English.) Another quarter could do basic addition but little more.[178] There is good reason for the new emphasis on education and on trying to motivate the young. We would not be competitive with societies where the young are still motivated to learn and to work, even if there were no wage differentials—and there are enormous wage differ-entials.

The President should heed Felix Rohatyn's question: what good is training if there are no jobs (p. 187)? Education and job training are essen-tial in the modern world labor market, but they are not a sufficient solu-tion. As we saw in chapter 11, even the skilled are losing jobs.

Educational reform will become a lot more practicable if there is a better fit between jobs and applicants. This leads back to the issues of migration and fertility.

The Benefits of Joblessness

There are now about 5 million people on Aid to Families with Dependent Children (AFDC)—almost all of them single mothers—plus about 10 million babies and children.[179] Welfare reform is supposed to get these mothers off the rolls, but from the women's perspective there are a lot of reasons to have a baby and become entitled: a monthly paycheck, access to Medicaid, and help with other welfare programs such as housing. In that world, there is not much incentive to go out into the economy—despite the President's exhortations—and very little they can do. The young women on AFDC are probably largely unemployable in this technological society. One estimate is that only one-fourth of them could pass the IQ test for the U.S. armed forces.[180]

An internal government study, quickly squelched, estimated that 2.3 million new jobs would be needed to accommodate welfare recipients pushed off the rolls after two years.[181] Whatever the figure, much of it will be make-work jobs, and this raises some social questions. Are the young mothers really better off with such an arrangement? Are their children better off if they must be put in creches so their mothers can work?

If the young women do get real work, they are probably competing with somebody. This is inescapable, but it doesn't make the unemployment problem any easier.

I would argue that the best solution is a slow one, and it requires restructuring welfare so that lower fertility is rewarded. An unidentified source in Clinton's welfare task force said that one of the ways to shrink the AFDC rolls is to mount "a campaign against teenage pregnancy."[182] A good idea, and quite compatible with Republican feelings. The Republican Right demands that welfare not reward young girls for having illegitimate children. Others respond that to discourage pregnancy is probably to encourage abortion. The provision was struck from the 1996 welfare legislation.) Whether or not they see their position as a demographic one, the effect would be to discourage childbearing, particularly among those least prepared to raise them, and that is a good idea. I have one warning: we must not penalize the children.

Welfare without Dishonor

I am not optimistic that we will address American fertility or make the decisions on immigration and on trade that would protect less skilled American workers. Ergo:

> *Unemployment is not a passing problem, and welfare must be geared to the long term, without destroying the people involved.*

Americans resist massive income transfers to welfare recipients and are not interested in a leveling of incomes. However, I hope that few people would really want their compatriots to be hungry or cold in the streets.

Is there a vision of a welfare program that can function within these perimeters?

■   ■   ■

My beat is population, not welfare. As I look at the connections, however, the rough outline of a workable welfare and unemployment policy seems to emerge.[183] It would have these elements:

1. Universality. Help should reach anybody who really needs it.

2. Modest expectations. The industrial world has experimented in the past several decades with the height of the "safety net." Some, like Sweden, almost obliterated the differential in living standards between those who work and those who don't. They are finding that it encourages idleness and that they cannot support the net at that level. We may need to experiment with simpler solutions, such as soup kitchens and the simplest of housing. Hong Kong in the 1950s helped to accommodate a flood of refugees from China in the simplest and least destructible of buildings with a central plumbing core and cubicles to live in.

3. Realism. Welfare and unemployment must be treated together as aspects of the same worsening problem. The street kids are not irrational; they play the cards as they see them, and their estimate of their prospects in the mainstream may be right. The young mothers are not competitive in this society, and they probably know it and hang on to what they have. The country is probably going to need a massive and continuing program of "make-work jobs"—some sort of national service corps, bigger and more diversified than the AmeriCorps program that President Clinton launched and the Republicans are trying to end.

4. Dignity. Dependency degrades people. There is much that needs to be done in this country, even if doing it cannot command a competitive wage. Look at our cities, or remember what the Civilian Conservation Corps (CCC) did in planting trees and improving our forests in the 1930s. Our parks, public lands and national forests could use help. Everybody who is not completely disabled should have a job if they want it, and participation should earn some remuneration. Responsibility begets responsibility. Young people would be less likely to trash urban parks if they had helped build them. The supervisors should be members of the service corps, not members of a separate bureaucracy. This would get rid of the handout mentality, and trainees who showed promise could work their way

up. Similarly, educational scholarships and training opportunities should be offered through the corps. Bringing in the upwardly mobile would remove the stigma, and the corps would be seen as a stage of life through which many successful people pass—and perhaps even as a period of real national service.

5. Simplicity. The present crazy contrivance of programs should be rationalized into something more equitable and understandable. It should not put the recipient at the mercy of the administrator. For example, if the public food and lodging arrangements were simple enough, they would not attract those who could afford better, and they could be offered without a means test. All those welfare workers chasing around trying to enforce the rule against a "man in the house" could be put to better uses. One demeaning aspect of AFDC would be eliminated, there would be fewer single-mother families, and kids would know their fathers. It would be good for the fathers, the children, the mothers and probably society itself.

6. Toughness. All this may sound a bit too much like an idyll. The Chinese did not trash those Hong Kong housing units—but the Chinese had more social self-discipline than is presently evident in our society. I have seen public housing made uninhabitable by its inhabitants. Street sleepers stay out of shelters for fear of what goes on inside. To make anything work in our demoralized society, we will need to become much less tolerant of antisocial behavior. There are laws, and we will need to enforce them, even if this requires the segregation of the incorrigible.

7. Generosity. Some but perhaps not all of the cost could come from the $250 billion or more spent on the existing hodgepodge of welfare, unemployment compensation and farm support programs, and perhaps even from the "corporate welfare" that gets slipped into so much legislation. We may need to look beyond those sources to Leontief's proposed social transfers, moving additional resources from those in the system to those who have fallen out of it. With a labor glut, skills are not being rewarded (figure 11-2), and the rich have profited. The times demand the opposite of the current enthusiasm for flat tax rates (p. 230).

Above all, consider our children. One of the primary functions of our social species is the social rearing of the young. That responsibility has been neglected as women have moved into the job market. It is not just welfare families that need help in making sure that young children get fed and exposed to the stimuli and experiences that start the learning process.

"Latchkey children" are a major national problem. Welfare mothers (and indeed fathers) could serve in supervised creches, learning to take care of their children and those of working people. Any new welfare legislation will spend billions caring for the children anyway. One cannot just abandon them in the effort to cut welfare; most of us would have a moral problem about casting aside helpless and often unwanted children. Enlightened self-interest points in the same direction. If they grow up ill-tended, badly fed and perhaps abused, we can be sure they will create trouble down the road.

Whether welfare is managed by Washington or the states is of secondary importance; the states will have to coordinate their programs, anyway, so drifters don't just drift to the most generous states.

If my proposal strikes you as an unwelcome addition to the government's role in America, consider trade and population policies that might in time diminish the problem. I say "diminish" advisedly. There is no way of knowing how far technological displacement will go and no models to suggest whether a stabilized, perhaps smaller population could be structured to provide enough jobs for all those who need them. All we can say is that smaller cohorts entering the labor market would be a lot more manageable than large ones, and this requires a population policy.

Immigration and Welfare

Immigration affects welfare in two different ways: first, it increases job competition and puts more people out of work; second, it raises welfare costs directly as immigrants go on welfare. Between 1983 and 1993, according to the GAO, immigrants' share of AFDC rose from 5.5% to 10.8%. Their proportion of Supplemental Security Income (SSI) recipients trebled from 3.9% to 11.9%. By one analysis, 20.7% of immigrant households receive one or more forms of welfare, compared with 14.1% of native households.[184]

Dramatic as that may be, the direct costs of immigration are secondary to the displacement effect and much less destructive of the social fabric. Refugees on relief may force taxes somewhat higher, but losing one's job is far more traumatic.

I will conclude this section by reminding readers of the first part of it. Without a policy on the forces that create welfare, there is no way that the government can simply get people off welfare without tremendous human costs. The President and Congress are not going to get very far with welfare reform until they look at the issue whole.

HEALTH AND HEALTH CARE REFORM

Health care reform has been a major national issue since the 1992 elections. The idea then, fueled by the upset victory of a universal health care advocate in Pennsylvania, was that the American people demanded it. Since then, faced with budgetary realities and the Republican sweep in 1994, the question has become a partisan issue. President Clinton in 1993 proposed a Health Security Act that alienated conservatives because of its complexity and cost.[185] As of this writing, the Congressional Republican majority wants changes in Medicare and Medicaid to control their rising costs but has not spelled out a conceptual alternative other than to pass the hot potato to the states.

Neither side in the debate has noticed the demographic context in which any health program must work. The Republicans seek to cut the rate at which health care costs grow, but remarkably enough, nobody has made the simple point that it is harder to control that growth with a growing population than with a stable one—and the country is growing 10% per decade.

We have something of a national habit of understating the difficulties when we want to do something. When President Johnson created Medicaid and Medicare, the country did not address either the demographics of a changing society or the costs of the proposed programs.[186] Demographic realities could wreck a health care program in several ways. The costs of health care could escalate beyond the nation's ability to pay them. We have an aging population structure, massive immigration, AIDS and the possibility of other plagues.

The other side of the coin is that a health care program will affect the rate of population growth, either accidentally or by choice. A liberal program would be a magnet to immigration unless the government finally bites the bullet and puts in place a system of identification that makes it possible to know who is entitled. Immigration policy and health policy must be considered together if we do not want to find ourselves serving the world. As to fertility: a health care plan will influence it in different ways. It would offer an opportunity to steer our demographic future if the nation were ready to start.

Finally, there is a connection between health care and unemployment. An attempt to force business to absorb health care costs will make the alternatives—automation and departure to other countries—more attractive.

Let me take up these connections in rough sequence.

Multiplication of the Elderly

The nation is headed for an expensive surprise if it estimates the public costs of any proposed health program on past experience rather than current trends. The following graph shows the anticipated growth of the elderly. Count on it. Only a plague will keep people from getting older.

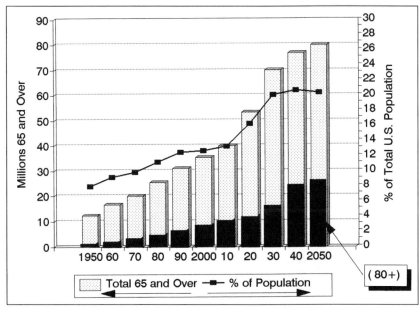

Fig. 12–2. U.S. Population, 65 and Over
(*Census Bureau 1993 Middle Projection*)

We must anticipate a particularly sharp rise in 2010–2030. It represents the last hurrah of the post–World War II baby boom. (Any enthusiast for population growth should consider that poor generation: going from crowded schools in the '70s to trouble finding jobs in the '80s, to the layoffs and deteriorating job structure of the '90s and finally to the role of an unappreciated social burden a few years hence.)

Look at the graph. That is quite a curve. The numbers of the "old old"—those 80 and over—will grow even more sharply and will peak a decade or so later. They are the ones who will require the most care.

Families used to support their old folks. The burden has been shifted to government as Medicaid and Medicare have provided an alternative care supplier. This means, one way or another, that the support comes out of taxes.

The cost of the President's plan would have escalated rapidly in decades to come. It offered prescriptions, long-term care and treatment for the diseases of the aged, which can be expensive. The plan's proponents said that long term care would "emphasize home- and community-based care." A laudable goal, but it did not say how we would pay for it.

The Republicans argue that we may not be able to afford unrestricted entitlements and that the government should preserve room to draw limits as to how much can be spent on each individual. They are right. Medical care now absorbs 14% of this country's GDP, probably the highest fraction in the world, and 42% or more of that is publicly financed. Entitlement programs have been a potent force driving our budget deficit in the past two decades because their size is dictated by demand (as defined in the legislation), rather than by the budgetary process. The nation will find that it cannot do all it would like for everybody, and it may need the flexibility to address the questions: What are the priorities? How can we best use each medical insurance dollar? Should we, for instance, try to underwrite extremely expensive operations for the elderly or procedures for others when the odds against full recovery are not very good? Oregon has led the way in addressing these questions, and to its credit the Clinton administration allowed it to proceed with the experiment.

Basic health care for those who cannot afford it seems a necessary goal in a civilized society, but every society has limits. As in many other areas, we have been reluctant to acknowledge those limits. Technology has made remarkable medical processes possible—at remarkable prices. We have come to feel an obligation to do everything possible, even if the result is to preserve life in people who have descended into mindlessness. None of the advances has altered the reality that people, like other animals, must die. The growing popularity of living wills suggests that a portion of our population recognizes that truth, but politicians dare not say it.

There is something of a panic right now as to how to meet the medical and welfare costs of that aging generation. Immigration proponents say we must bring in young immigrants to support them. Natalists argue for more children. Both miss a central point. The problem is a transient one. A rise in the elderly population is natural and indeed inevitable. The proportion inevitably will rise as any nation gets off the population growth treadmill, unless population growth is stopped by hunger or epidemic. There are only two ways to keep a population young: high mortality or a constantly expanding population. Africa has both, and they are not a solution.

Eventually, we must stabilize. To do so we must adjust to a changed age structure, which is not necessarily a problem.[187] There will be fewer children to support and more aged. Most people pay for much of their children's costs, so it will become a tradeoff of lower direct costs for higher taxes. Moreover, the rise in productivity described in chapter 5 means that we won't need all those young people to support the old—we will need productive workers who pay their taxes.

The "aged boom" is much less serious a prospective problem here than in Europe (chapter 7), because we have not had a period of extremely low fertility. Even if U.S. fertility dropped to 1.5 in the next generation (i.e., the "two-child family" proposed in chapter 13), our dependency ratio in 2050—after the baby boom had passed on—would be 63%, which is a good ratio reflecting a healthy society with moderate fertility and long life expectancy.

AIDS

The politicians are going to hate to have to face this one. AIDS is going to eat up more public health money unless we begin to treat it seriously as an epidemic. Reported AIDS cases have risen steeply, from 4442 in 1984 to a cumulative total of 245,000 by 1992, with probably more than a million people probably HIV positive. Infection rates seem to be declining among gay White men 25–44 years old, but rising among younger gays, drug users and—particularly—women.[188]

Perhaps even more ominous, the epidemic in the United States is growing fastest among the young. One expert reported that AIDS cases among adolescents had risen 77% in two years.[189]

AIDS is a particularly insidious epidemic, partly because of the ten-year average incubation period. Barring a cure that is not yet in sight, we may anticipate an explosive growth of an expensive class of potential health care recipients. Like care for the elderly, this expense will probably tend to fall in the public sector of the health program; employers will do everything they can to avoid it.

Sympathizing with the victims, putting more and more public money into treating them, and demanding more federal expenditure on the search for a cure is not necessarily a sufficient approach to this massive threat.

As the World Health Organization has pointed out, AIDS is in large part a venereal disease. If AIDS expands into a major plague, it will be the first one that ever happened because the victims (other than the babies born with it) knew the cause but didn't want to change their habits and the health control authorities were not permitted to take a realistic approach to controlling it. That formulation makes it less glamorous. To put it in perspective, Can you imagine the idea of benefit balls, donation drives, glossy brochures laced with famous names, and appeals to the President, to help syphilis sufferers? To be sure, the moral fervor is connected with the effort to assert the legitimacy of homosexuality. This should not obscure the need to treat AIDS as a public health problem. Mandatory testing of prostitutes, people in the armed forces and those applying for marriage licenses used to help to control venereal disease. AIDS is a much more malignant threat, and the health experts should be asked what should be done—from a strictly public health standpoint—to control it, or we will watch a ballooning of the costs of health care.

Plagues

The popular press seems largely to have missed the story, but there has been a fundamental erosion of confidence within the scientific community as to whether we can handle other infectious diseases. Humankind is being subjected, in a sense, to a counterattack in a war that we thought we had won. We are watching a multiplication of threatening disease strains. One source lists 28 new diseases and 14 that have reemerged, including old favorites such as cholera, malaria and tuberculosis (which has risen dramatically in the past decade, now kills 3 million people a year and is called a "global crisis" by the WHO[190]). The pathogens are learning to handle the drugs and antibiotics that we launched against them two generations ago.

In the United States, the death rate from infectious diseases rose 58% from 1980 to 1992 and would have risen even if AIDS deaths had not been counted. Dr. Joshua Lederberg, Nobel laureate and President emeritus of Rockefeller University, says we are in a "race with microbial infections."[191]

We have been warned by the editors of *Science* magazine of potential new epidemics much more frightening than AIDS. The National Academy of Medicine has issued a warning of the possibility of pandemics rivaling the Black Death, which killed perhaps one-fourth of humanity; as a worst case "it is conceivable that pandemics could do so again." The Academy attributes the threat to a breakdown of public health, changing land use, international travel and trade, population change, new technologies and industries and microbial adaptation. The American Association of Microbiology warns of rising rates of virus infection, genetic mutations and pathogens spreading by new pathways.[192] This is not just flashy journalism; these are the experts.

What do we learn from this gloomy recital? First, the ability to prevent epidemics may suddenly be much more important than national health programs. Investment in the Centers for Disease Control and

Prevention may be more critical than insurance coverage for expensive and marginal health services.

Second, the epidemiologists are confirming the importance of any measures that limit population growth and the resultant breakdown of public health anywhere in the world.

Third, immigration affects public health. The new pathogens are originating largely in the third world (chapters 1 and 7). It would be unfair to blame immigration for all potential epidemics. Pathogens can find other ways to jump borders, or they can mutate, as the hantavirus apparently did in the United States. Immigration did, however, introduce new strains of resistant tuberculosis to the United States, and cholera has appeared here, which suggests the need at least for better screening of immigrants. We need better control of illegal immigration and better ways of identifying who is here.

The Magnet Effect

The AIDS epidemic, with the possibility of others to come, raises other questions about immigration policy.

AIDS cases worldwide may be approaching 5 million, with no end in sight (chapter 7). In Haiti, 7% to 10% of adults have HIV, the forerunner of AIDS. Haitians with HIV have been brought into the United States by court order. Congress has voted resoundingly to forbid the immigration of HIV-infected persons, and President Clinton signed the bill despite his campaign statements in favor of admission. It is far from certain that the decision will stop the movement. The availability of free hospital coverage for AIDS and other plagues will attract sufferers in less generous lands and, given the casual way we enforce our immigration laws, many of them will probably be able to get here.

Perhaps it sounds heartless. One would like to be able to help everybody, everywhere. There is, however, a conflict between moral obligations, driven by the reality that resources are not infinite. We may well have trouble in doing what we would like for our own people in the years ahead. I hope that chapter 7 put to rest the misapprehension that third world population growth, the chief engine of migration to the United States, is waning.

The question is not just whether health reform will attract very expensive immigrants such as those with HIV. Universal health care would encourage immigration and thus population growth. Added to food stamps and other welfare, health care would be a strong inducement to come here or, once arrived, to stay. Already, the practice of crossing the border for treatment in welfare hospitals is well established. One illegal alien who had just lost his job told a reporter, "It's still better than where I came from." Others may agree.

The pendulum has swung. Republicans say most benefits should be available only to citizens. The Democrats would add only legal residents. Housing and Urban Development (HUD) Secretary Henry Cisneros said we should deny health and welfare benefits to illegal aliens[193]—or, I assume, to nonimmigrant visitors, of whom there are about 20 million every year. Presumably, both groups would continue to get treatment for contagious illness and emergency care, for humanitarian reasons and the protection of others.

The question is whether the nation can enforce that decision.

A Question of Identity

We will need to know who is entitled, or we will find ourselves offering the health program to the whole world. That in turn will require a better system of identification, based perhaps on one's Social Security number and a call-in system such as has been perfected by credit card companies.

In one of several turns on the subject, President Clinton said the unutterable: that his administration is studying the feasibility of such a card to help manage the health care system, even while remarking that a national identity card "sort of smacks of Big Brotherism."[194] Apparently, he recognized that we must be able to identify people if we are to have an affordable national health program. When the crunch came in 1996, though, he wasn't heard from. The proposal for better identification has been around for a generation (see chapter 13). It is important for any national health program; it is central to any hope of controlling illegal immigration, which is an issue far transcending health alone. The proposal arose in Congress again in 1996. We will see in chapter 17 that business lobbying and political equivocation have again pushed it off into an indefinite future.

The Impact on Fertility

A true national health program would constitute a fundamental change of direction, akin to the New Deal. It deserves policy debate integrating it with other broad national questions. Where, for instance, do we want to go, demographically, and how does health care relate to that question?

As a beginning, erstwhile Surgeon General Joycelyn Elders pointed out in sulphurous language that the welfare problem would be smaller if Medicaid funded family planning to avoid unwanted children.[195] So would population growth.

By extension, perhaps it would be unwise to make health insurance too generous in covering pregnancies. We could, for instance, look to the

Singapore example and offer maternity benefits only for the first two children. A priori, one assumes that women's decisions whether to have babies will be somewhat influenced by whether they have to pay for them. If the nation wants to avoid subsidizing irresponsible pregnancies—as many Republicans in Congress now argue—it must shape its health and welfare programs accordingly. It is also true, if still unsaid, that that would slow U.S. population growth.

Some would charge that such a policy is intolerable governmental interference in private behavior. I would counter that behavior is no longer private when it is publicly financed and has major undesirable consequences.

We see ourselves as a modern nation capable of shaping our future. Let me point out that Iran, which U.S. opinion sees as a backward theocracy, has passed legislation to withhold insurance benefits and maternity leave for children after the third.[196]

Employment

President Clinton proposed to charge employers for much of his proposed health program. Employers responded by shifting to temporary hires. They avoided commitments to their employees' health and retirement, but in the process they made the problem of unemployment worse, and more people lost their health insurance.

As costs rise, businesses will automate or move their operations out of the country, both of which make employment prospects in this country bleaker. As if it were fated to propose policies with synergistic ill effects, the Clinton administration's sponsorship of NAFTA made it easier for American business to move its operations to Mexico (p. 187). Health policy cannot be separated from trade and employment policies.

The Population Solution

WE TRY TO SOLVE OUR PROBLEMS without addressing the demand side. We worry about our collapsing cities and the drift of our society toward a two-tiered stratification of a few "ins" and more "outs" but don't ask "why?" We seek to reduce joblessness without addressing the growth in the working-age population. We endanger our agricultural base by encouraging higher yields to accommodate population growth. We seek technical fixes for pollution without considering whether we should stop the growth that drives the increase. We seek to control welfare and health spending while passing immigration bills that raise their cost. Seen from a reasonably detached viewpoint—from Mars, let us say—the arguments for arresting U.S. population growth would seem so compelling as to raise the question, Why isn't it being done?

Unlike most policy fixes, a demographic policy would begin immediately to reap budgetary savings. Yet very few of our politicians and pundits even consider the idea. Proposals to address population growth find few takers because they conjure up fears of China's one-child families, of fertility controlled by forced sterilization and mandatory abortions and of a "Fortress America" with all migration barred.

The reality is much simpler and less frightening. A turnaround could be achieved with the "two-child family" (see below) and with immigration returned to the levels that prevailed for much of this century. On the assumption that action is held back by unwarranted fears, let me show the relatively gentle adjustments that could turn U.S. population growth around. If the will were there, the solution would not be difficult.

There are no points given for being "in favor of" stopping or reducing the growth. The question is, What do we propose to *do* about it? We need the courage to take on our demographic future. If we can so confidently urge African nations to face their demographic problems, what about doing it ourselves?

STOPPING THE TREADMILL

There are only two variables available to influence U.S. population growth: fertility, and migration. Nobody wants to raise mortality. If we moved instantly to the two–child family and concurrently brought immigration under control at a level of 200,000—which was sufficient to absorb many legitimate refugees from World War II—our population in the next century would decline to the level shown in figure 13-1 below.

If parents indeed stopped at two children, it would lead to a total fertility rate (TFR) of about 1.5 children because not all women have children and some have only one.[197] In all the debate about population growth, this simple reality is almost universally ignored. A national TFR of 1.5 in turn would make possible a gradual turnaround in the population growth that presently drives our social and environmental problems.

That combination offers the prospect of addressing our own future without slamming the door.

The black bars labeled "planned" represent the population curve that would result if the two–child family and net annual immigration of

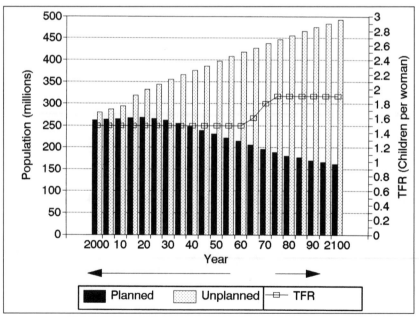

Fig. 13-1. The Two–Child Family. Fertility and Population
200,000 Annual Immigration

200,000 were to be achieved instantly. The "unplanned" bars are for comparison and show where our population is now heading.[198]

The "planned" bars are unrealistic, since things do not change instantly. They do, however, make the point that population could be brought down to 150 million and stabilized there, with fairly modest changes in behavior and policy. Perhaps even more dramatic is the TFR line (right scale). By the middle of the next century, families could be encouraged to have more children (for a TFR of 1.9) if as a society we decide that the 150 million range is a good size for the nation. Some countries have actually gone through a comparable experience. Singapore was so successful in bringing fertility down that it has changed course and encouraged somewhat larger families. Japan and Taiwan are making the same shift (perhaps prematurely).

All demographic projections are just that: projections, not predictions. If, less optimistically, we assume a slow decline in fertility to 1.5 by 2050 and the same immigration, the population in 2100 will be 193 million and declining.[199] The difference between the two projections is one of timing; the more gradual projection simply defers the population turnaround.

Another graph could be constructed with lower fertility and higher immigration and arrive at the same population size. Or, as a compromise, immigration in the 500,000 range and the two–child family would lead to the end of U.S. population growth in about 2040, and by 2100 population would be down to 233 million, of which nearly 30% would be post–2000 immigrants and their descendants.[200] The point is that, with a clear view of where we want to go, we could decide the ideal mix of fertility and immigration to get there.

THE TWO-CHILD FAMILY

Migration is the more important issue right now in order to stop the growth; fertility must be addressed if we are to turn it around. It may be clearer, however, for me to start with fertility.

Fertility: "Stop at Two."

The idea seems so transparently simple. Now that child mortality is under control, parents should have two children to carry on the family. It may once have been necessary to have many children so that some might survive to maturity, but that is no longer true. To behave as if we still lived in that unhappy condition is to invite population growth that goes within a very few generations from intolerable to unimaginable to mathematically absurd.

The two-child family is a more attractive idea than the Chinese goal of one child. We can afford that luxury because we are not quite so far down the road of overpopulation. Two children avoid the loneliness of the only child. (With the one-child family, there are no siblings and no cousins, and four grandparents share only one grandchild to spoil.) But if there are no more than two, parents and society can hope to provide them with a decent upbringing and—even at today's educational costs—an education to start them off in life. Fewer families would fall below the poverty line. Fewer would need welfare.

The New American Tradition

When people were introduced to the mother of a dozen children, the traditional response was to offer congratulations. By now, it is as likely to be incredulity or embarrassment. "Stopping at two" is already the norm for most Americans. Of mothers completing their childbearing years, 70% had only one or two children. Married women's expectations have dropped sharply in the past generation, from three or more children to an average of just over two.[201]

To achieve a population turnaround, substantially everybody must stop at two. If third and fourth children are more than an exception, the demographic prospects will be darker. There must be an explicit understanding that going beyond two is socially undesirable.

If this sounds like too much management for the American taste, I would suggest that the reader look again at the "unplanned" alternative in the graph and consider whether it is worth avoiding. The nation has learned how to manage mortality—at least for the time being—without a corresponding willingness to manage fertility. The resultant demographic imbalance lies at the heart of our environmental and social problems. It leads to the painful necessity to choose between free choice and a population policy, between limiting immigration and acquiescing in a continuing demographic explosion in the United States.

Stop Preaching to the Choir

Most population literature is just that: written literature. Readers tend to be educated, but the educated generally are not fertile. The population argument must reach the majority who get their information in other ways. I don't know just how, but America is full of publicists, advertising specialists, public relations types, "spin doctors" and other practitioners of the black arts of public persuasion. We need to seek their advice. The fertile will stay fertile unless they are convinced that limiting their own fertility is a good thing.

We face the task of changing values. In good part, the problem is simply one of getting people to focus on the fact they have a choice. People, just ordinary people, usually have a very clear picture of their self-interest. One can make the argument about the common good but—particularly in an age with little sense of community and much hostility between groups—one needs also to appeal to self-interest. Let us consider how that might be done.

Incentives and Disincentives

We must find ways of reaching people where their interests lie. We must—with compassion and caution as to the potential dangers—use incentives and disincentives to give force to population policy as we do for other national policies.

The population argument starts as an abstraction. Abstractions tend to be the province of the educated. By and large, babies are not the product of such calculations. Among the poor young girls having illegitimate children, pregnancies happen because of hormonal drives and group pressures. As with her choice of sneakers, the girl may want to be "in"; she may want to keep her boyfriend even if it means giving in to his macho desire to father a child. Perhaps the girl simply wants something of "her own." The babies thus born are the most vulnerable. Somehow, those kids need to decide on their own to say no.

The nation creates incentives and disincentives all the time, unconsciously or to achieve social objectives (wise or unwise). Subsidies, agricultural price supports, tax policies. We are entitled to use them to influence choices about fertility. The measures could involve changes such as the following ones:

- Welfare—structure it to discourage pregnancy, as the Republican majority in the House proposes.
- Health insurance, as I suggested above.
- Vacation and maternity leave policies.
- Public housing—limit the space to make it crowded for more than the nuclear family of four.
- Access to the best schools for the first two children, as Singapore allowed.
- Tax policy—deductions or credits only for the first two, unlike the present proposals in the United States.

Above all, we must assure that young women recognize and can enjoy the benefits of delayed pregnancy and limited children and have the sense of self-worth to pursue them. There are various ways of encouraging lower fertility.202 I will not detail them here. The most difficult problem is how to pursue them without penalizing the children beyond the second who are born anyway (chapter 12).

A Question of Fairness

Any policy of discouraging high fertility must be across the board. It cannot be targeted simply at the poor. The two-child family must be a goal for everybody. Any mention of a population policy is met with fears that it is targeted at an economic class or racial group. I think that needs to be discussed. Let me start with a graph.203

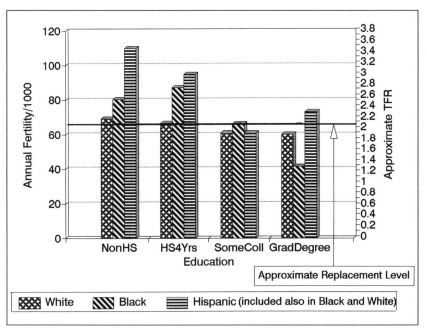

Fig. 13-2. Fertility of U.S. Women. Ages 15–44; by Race and Education
(*Census Bureau, Fertility of American Women, 1988, '90, '92 and '94*)

Fertility is related at least as much to educational level as to race. A graph of fertility plotted against income would show a similar correlation. If we had sufficient data to plot Asians on this graph, I suspect they would show the same correlation. Generally, the more educated women are at or below replacement level fertility, which would eventually stop population growth if there were no immigration.

It is not an insult to the uneducated to urge the adoption throughout society of an approach to fertility that will improve the condition of all,

and perhaps of the uneducated particularly. Nor is it racist. In total numbers, more Whites than others would be asked to reduce their fertility.

There is a phenomenon the demographers call "shifting shares." Over
time, the more fertile segments of the population will become a larger
proportion of the total and will thus increase total fertility and population growth. (This rests on two assumptions: that higher mortality does
not counteract higher fertility, and that succeeding generations follow the
patterns of their parents.) In the UN constant fertility projection in figure
1-2, worldwide total fertility would rise from 3.38 in 1985–1990 to 5.7 in
2145– 2150, because more of the population would be descendants of the
more fertile. This is the human equivalent of Darwinian natural selection.[204] The importance for our present discussion is that any effort at
population limitation is bound to fail if it does not reach substantially all
segments of the population. Voluntary participation by the awakened is
not enough.

There does seem to be a cultural propensity among Hispanics to have
larger families. Since they are 10% of the population, and rising fast, their
fertility could undermine any effort to stop or reverse population growth.
They tend to be family oriented, and that is probably a factor in their
higher fertility. One can hardly find fault with a strong sense of family in
a society that is rapidly losing it; the problem is to disengage it from high
fertility. There are some reasons for hope. Spain itself, the prototypical
Hispanic culture, has the lowest fertility in the world, along with Italy.
Since the Roman Catholic church plays an important role in Hispanic
immigrants' lives, Spanish Catholic lay leaders could perhaps help in
reaching these immigrants. For example, could American population
organizations sponsor speaking tours and talk shows by appropriate
visiting Spanish cultural leaders, perhaps on Spanish–language radio
stations?

Hope for a change of attitude within the Vatican seems perpetually to
recede. Pope John Paul II opposes even the advocacy of small families. He
has condemned "propaganda and misinformation directed at persuading
couples that they must limit their families to one or two children."[205] Perhaps some day . . .

IMMIGRATION IN MODERATION

The two–child family alone will not get us there. I have pointed out
that 43% of U.S. population growth in this century has been post-1990
immigrants and their descendants, and post-2000 immigrants and their

descendants will probably cause 91% of our growth in the next century (chapter 9).

A position "in favor of" halting population growth without a policy on immigration is simply a nullity.

In the last three decades, migration to the United States has trebled, and we don't really know the whole story. Communities of migrants from many countries have developed in the United States. The existence of such groups fosters "chain migration," and third world population growth promotes it (chapters 7 and 8).

The average annual level of recorded immigration between the passage of the Immigration Acts of 1924 and 1965 was 198,000. If we were willing to bring net immigration back to about 200,000, our demographic future would improve dramatically.

How could we do it?

Slowing immigration is politically difficult but conceptually manageable. There is no problem convincing the poor—even those minorities that are themselves mostly recent migrants—that immigration hurts them because it lowers wages and makes jobs scarce. That is why polls find those groups in favor of limiting immigration (p. 261). The problem is that politically powerful interests want high immigration (chapter 15).

"Doing something about" immigration requires changing the Immigration Act of 1965, which led to chain migration as new immigrants bring their extended families to the United States. The Refugee Act of 1980 needs a thorough new look. The Immigration Reform and Control Act of 1986 (IRCA) was effectively gutted by making it almost impossible to know who is here illegally, and it has led to widespread fraud. It needs to be revisited. There is no effective control over immigration by those who overstay their nonimmigrant visas. The Immigration Act of 1990 led to a 40% increase in annual immigration, and probably more because of the chain-migration effect. It should be undone. The ideal: a new law and better enforcement, bringing net immigration down to about 200,000.

Immigration reform will require change in several ways:

- *Turn off the magnet.* People come here for jobs. We need rules against not just the hiring but the continuing employment of illegal immigrants to get at one root of the U.S. unemployment problem (chapters 11 and 12). We could make it less attractive for illegal entrants or visa overstayers to stay here by limiting the issuance of professional and permanent drivers' licenses to legal residents.

- *Develop reliable means of establishing identity.* If we cannot identify illegal immigrants within the country, we can control immigration only at the border, and that would require a "fortress America" mentality and controls on movement that we would rather avoid. With a better way of identifying people, we could afford a less rigorous control of the border in the expectation that we could identify and deport those who overstayed or got through the screen.

Various groups, including Congressional commissions, have urged for years that there be a better process of identifying U.S. citizens and aliens with a right to be here. The IRCA bill initially provided for a better system of identification, but as part of the final compromise, that proposal was weakened to nullity, and since then there has been a proliferation of fraudulent documentation.

There are other powerful arguments for better identification, such as the protection of wage and labor standards, the collection of taxes from all who should pay, and the identification of terrorists and criminals. Any national health plan requires it (p. 201).

The central issue is that with a better way of identifying people, the INS could identify who is not entitled to be in the country and thus for the first time could effectively enforce our immigration laws.

The idea has some support in both parties and at different levels of government. The Commission on Immigration Reform (the "Jordan Commission," after its late chair, former Congresswoman Barbara Jordan, a Democrat) came out for it in 1995. The states are beginning to demand a better identification system. California is the immigrants' primary destination, and it is foundering. When Republican Governor Wilson was a senator, he sponsored the biggest loophole in the 1986 IRCA and later voted for the 1990 act increasing legal immigration. He has seen the light. He has called for a series of measures to bring illegal immigration under control. With considerable fanfare—he appeared at one news conference surrounded by stacks of forged identity cards—he asked President Clinton to "use California as a testing ground for a tamper-proof identification card to combat illegal immigration."[206]

There is strong opposition to a "national ID card," and it is encouraged by businesses that don't want to lose their freedom to employ illegal aliens. In fact, we already have a national identity card, the Social Security card, which is used as an identifier for taxation, drivers' licenses, and other purposes. We do not control it very well or use it very effectively. We must have a better system.

- *Control the border.* Whenever the immigration issue heats up, critics of the government demand that it "beef up the Border Patrol," and the government announces how tough it is getting. Authorization bills are passed and then frequently forgotten in the actual appropriation process. Immigration will not long be hindered by the tragicomic game of Border Patrol agents chasing would-be immigrants. Particularly in light of continuing turbulence in Mexico, we should plan to use the military as necessary, not to supplant the Border Patrol but to supplement it. This proposal usually elicits a nervous response and references to "the Berlin Wall." It is not such a departure as it is perceived to be. Safeguarding borders is a normal function of armies. The Italian army is doing it right now in Calabria. We have used the Navy and Coast Guard to fend off illegal arrivals by sea and the National Guard to help fight drug smuggling, and the INS has recently revealed that some 200 military officers are assisting it in technical capacities such as teaching the Border Patrol the use of sophisticated detection devices.[207] It should be doing much more.

■ ■ ■

There are more detailed recommendations available of ways to control illegal migration.[208] I would underline one cautionary note: our country must find a way to control immigration without cutting ourselves off from the rest of the world. In a world grown small, we cannot control the movement of ideas nor should we want to. We are coming to realize that all humans are inhabitants of one planet and that we must cooperate with others to protect it. Our country does not foster that cooperation if we raise walls around ourselves.

This suggests that we should strive to make it easy to visit but hard for visitors to overstay their welcome. If we can establish a system of identifying people, through reform of our vital statistics and the use of identifying documents such as the Social Security card, then we do not need to try to make our borders impregnable. Instead, we can enjoy the benefits of movement by welcoming visitors, secure in the knowledge that we can keep them from using their visitor visas to immigrate. We cannot do that now.

FOREIGN ASSISTANCE

The central argument of this book is that population growth is the crosscutting issue generating the tensions and dangers of today's world. It follows that

support for programs to arrest population growth should be the first priority of U.S. foreign assistance, not a minor and uncertain add-on.

This is the one area where foreign assistance of politically realistic proportions could have a worldwide effect. We cannot solve others' population problems, but we can help. It is not enough to hope for modernization to bring fertility down; some countries have done it, but much of the third world does not have that option. Countries that succeed in bringing fertility down with leadership and population programs, in advance of modernization, have a chance of escaping the African dilemma (chapter 7) and using their resources for bettering their people's lives, rather than chasing population growth. On that critical point, foreign aid can be important.

We have an altruistic interest in helping that process. We also have an enlightened self-interest. We would help ourselves by helping others to escape the pressures that lead to migration and epidemics originating in the vast unsewered slums of the poorer countries. Population stabilization would mitigate the conditions that generate instability, and that is in our interest. Eventually, it would ease the competitive pressures from low-wage countries that lead to unemployment and wage declines in the United States (chapters 8 and 11). A world without those tensions would permit more open trade and movement; it would be a better place for this country and others.

The Clinton administration has taken the international population issue seriously. State Department Under Secretary Tim Wirth says that "If we can't stabilize the world's population, we're not going to be able to control any other problems." Population assistance doubled in the first years of the Clinton administration, and the 1995 budget included nearly $600 million for population programs. A good start but still only about 4% of all foreign aid.[209]

All foreign aid is now under attack, and particularly population assistance, which has become linked in the Congressional mind with abortion. The Republicans, as of this writing, were somewhat divided, with the more extreme among them pressing for an end to foreign aid. This faction includes Senator Jesse Helms, who heads the Senate Foreign Relations Committee. Senator Mitch McConnell (Republican-KY), chairman of the

Senate Appropriations Foreign Operations Subcommittee, wants to zero out all population assistance and all aid to Africa.[210] The born-again Republican freshmen elected to the House of Representatives in 1994 agree. The tortured compromise was to put $356 million in the FY1996 budget but under conditions that will make it possible to spend only $72 million in FY1996.

The real priorities are seen in the numbers, and for years nearly half of foreign aid money has gone to Israel and Egypt. I am not optimistic that domestic political realities will allow those priorities to be changed. Former Senator Robert Dole and others are not necessarily against population aid but they are cultivating the domestic Israel lobby. The 1996 budget proposal emerged from Congress with reduced foreign aid, but with a protected $5.1 billion for Israel and Egypt, which forces a deeper cut in other programs. (This is a good example of Congressional responsiveness to money rather than to public opinion [see chapter 11]. A University of Maryland poll listed ten types of foreign aid and found that aid to Israel and Egypt is the least popular segment of our unpopular foreign aid program, with 56% of those polled favoring its reduction or termination and only 42% favoring retention or expansion. On the other hand, 74% wanted to maintain or increase family planning assistance.[211] Israel is, nevertheless, the one untouchable section of foreign aid.)

In an unpropitious political season, I would endorse Secretary Wirth's opinion and take it another step:

make family planning and related programs the top priorities
for foreign aid.

The criticism of foreign aid has some basis. Diffusion of effort over the years, misdirected capital projects and inattention to environmental consequences have all generated doubts as to how much we have really done. Economic progress is probably a product of motivation and institutional arrangements more than of foreign capital injection. I watched Taiwan taking off on its remarkable economic growth in the late 1950s. I came away convinced that the real U.S. contribution was not the money we supplied but the protection our presence gave to certain economic modernizers in Taiwan who otherwise would have been overruled by the military, which was primitive in its economic thinking and did not welcome painful economic reforms.

Population assistance can have a similar function. It is the central crosscutting issue, and the U.S. Agency for International Development has shown that it can focus and monitor its population efforts. U.S. participation can kick-start population programs where they languish. It can give impetus to the drive for lower fertility. It can support local people who are pressing for more active programs.

We are no longer in a position to fill investment gaps abroad. We could fill much of the unmet demand for family planning at a cost far less than the United States spends now on foreign assistance. The total world "unmet" need for family planning assistance (i.e., making sure that family planning is available to those who would use it) was estimated by a broad group of experts under UN auspices. Total funding requirements were estimated—liberally—at $17 billion dollars in 2000, of which international sources would (they hope) provide $5.7 billion. If the United States chose to lead the way in providing funding, one-third of $5.7 billion is less than $2 billion, which in turn is less than one-third of budgeted U.S. international assistance, even after taking out the $5.1 billion for Israel and Egypt.[212]

Alone, we can hardly solve the world's ills. We can help others help themselves. Eventually, their success will help us. It will help to create a world in which the infertile rich do not have to try to cloister themselves away from the fertile poor. U.S. cooperation with the third world would become much easier—and we have tasks ahead that demand cooperation, such as trying to save the Earth's environment.

The "Statement on World Population Stabilization by World Leaders" (p. 150) makes clear that most third world leaders would welcome such aid. They, more than we, see the costs of population growth. Population momentum is a frightening thing. Look again at that African population pyramid in figure 7-1a. We don't need to convince them so much as to help them. It is immensely important that we not delay.

The U.S. Government has redefined population aid as support for women's issues—education, jobs and rights such as property ownership (chapter 17). Those are legitimate goals in themselves but, as I will detail in chapter 16, they have a somewhat uncertain and tangential relation to population growth and they divert money from family planning.

I believe that the U.S. Government should once again emphasize the sponsorship of direct family planning programs to provide knowledge, support program leader and facilitate access to contraceptives. Within our present foreign aid expenditures, we could go a long way to assure that there are no unwanted babies, and that is a step toward stopping population growth. In addressing women's status, the United States can afford to play only a marginal role at best.

I would propose a second focal point for our foreign aid once we have provided what population assistance the recipients can manage. Let us not forget the extraordinary pressure that the "emerging" countries will be putting on the Earth's environment (chapter 7). They expect the industrial countries to pay for any measures to ameliorate those effects (chapter 8)—more than the industrial world is likely to think it can afford. Some help, however, might pay off in terms of changed third world attitudes toward pollution controls. Pollution hurts the polluters first. A key form of assistance would be to cover the licensing costs for benign technologies. I think here of China's coal consumption and of the clear warnings from the IPCC of the urgent need to move to ultraclean coal technologies everywhere. Think it over. It would be a manageable but perhaps vitally important contribution to a sustainable world.

CHAPTER 14

The War of the Paradigms

WHY DON'T WE ADOPT some of those population policies? The United States is the third largest nation and the largest consumer and polluter. We encourage others to deal with their population problems and yet cannot address our own. Why?

Each of us, consciously or unconsciously, carries a set of beliefs and mental images of the nature of reality—paradigms—that shape our reactions to most social issues. Anybody who has lectured on the population issue will, I believe, recognize a mind-set that challenges us: a deep faith in growth, rooted in the belief that the Earth was created for mankind and that God wills that we should use and enjoy it.

That belief argues against the importance or even legitimacy of population policies. Let me, in an admittedly speculative chapter, try to describe and rebut it so as to clear the way to a consensus that population growth is an issue.

THE FAITH IN GROWTH

Our Anthropocentric Heritage

The belief in human "exceptionalism" and the sense that the rest of nature exists to serve us go back to biblical roots, reinforced by the American experience. The Judeo-Christian tradition has not served us well in learning to live on a finite Earth. One short biblical passage has had profound repercussions down the years: "And God blessed them, and God said unto them, be fruitful and multiply, and replenish the earth, and subdue it: and have dominion over the fish of the sea, and over the fowl of the air, and upon every living thing that moveth upon the earth" (Genesis 1:28). That injunction still rings from the pulpit and in the minds of our contemporaries:

- Orville Freeman, environmentalist, futurist, church leader and former U.S. Secretary of Agriculture: "After all, the earth's resources were put in place to meet humanity's needs."[213]

- President Clinton: We must "keep faith with those who left the earth to us."[214]

- Pope John Paul II (even while promoting environmentalism): "God made him [man] in his image and likeness and gave him the earth as his inheritance, . . . Did the Creator not perhaps give the earth to men and nations so that it could be watched over and cared for?"[215]

It is a vision of the universe created for mankind to use, with the Earth at its center and a God that shares our image. Man the self-centered. Classical Greece, with its all-too-human gods philandering with humans, shared that ethnocentric image. This expansive self-image was perhaps tolerable until humans suddenly exploded in numbers and economic activity to the point of making good on the mandate in Genesis.

Environmentalists in southwestern New Mexico have been trying to restrain ranchers from misusing public land in the Aldo Leopold Wilderness. The ranchers call them "environmental pagans." By their reasoning, the ranchers have a point. Wasn't the land put there for them?

Thomas Gray, in his *Elegy Written in a Country Churchyard*, wrote the lines

Full many a flower is doomed to bloom unseen and waste its sweetness on the desert air.

The poem was meant to be a homily against pride, but those lines have always impressed me as an unintended monument to it. I would argue that the flowers didn't give a damn; they were programmed to attract a bug, not a poet.

One can look at Christianity in different ways. The passage from Genesis includes the injunction to "replenish" the Earth. St. Francis of Assisi, with his closeness to other animals and his doctrine of simplicity and service, may have been trying to warn his fellows of the dangers of self-importance. Christian theologians have offered rationales for reconciling Christianity with a respect for the biosphere.[216]

A Comfortable Belief

Growth, like the Juggernaut, is powered by people who believe in it, right or wrong. This book is full of examples of the ingrained tendency to see economic growth as a solution for problems.

The faith in unlimited scope for human opportunities was strengthened in the Romantic Revolution in nineteenth century Europe, when the

Industrial Revolution got under way and civilization burst its constraints. It was a congenial mind-set for America, with our frontier tradition, our vast physical size and a national myth originating in rebellion against restriction. The American Revolution was ignited partly by the Crown's efforts to hem us in east of the Appalachians. With very limited setbacks, our history has been one of growth and rising prosperity. Economic growth—worshipped under the name of GNP or GDP—is the popular measure of success (p. 146), simplistic though that may be. Human population growth is seen as strength—though why people can believe that is a mystery to me, in the face of the evidence provided by countries like Bangladesh.

Growth and Profits

That congenial faith is reinforced by self-interest. Growth is a tenacious belief because it does work—for a time. Ask realtors, developers, merchants, manufacturers, builders, well drillers, electricians, carpenters, landscapers, insurance agents, hairdressers, masseurs and most of the businesses in the Yellow Pages. It brings in business and with it prosperity. It also drives up taxes, drives down wages, fouls the environment, raises the cost of a place to live and generates rising tensions, but the immediate result is profit. *Carpe diem.*

Armed with the faith, believers simply dismiss warnings about population growth with the remark that "somebody named Malthus long ago thought population growth was bad, but he was proven wrong." Think positive. Ignore those who tell you that the good times cannot roll on forever.

"Growth Helps the Environment"

Growth advocates have a rationale to reconcile their belief with environmentalism. They argue that only with growth can we afford to protect the environment. Economic growth is thus proposed as the solution to the very problems that it has generated.

I think the proponents have confused growth with prosperity. If the growth is in health and environmental measures or in the introduction of more benign technologies, fine. But growth usually means higher consumption levels, more energy use, more driving and bigger houses. These in turn multiply problems such as energy production and climate change, acid rain and solid waste disposal. There are uses for prosperity, but the real problem is the human impact on the Earth. The first task is reducing the sources of damage; cleaning up the damage is a necessary task when that fails, and it helps if the funds are available.

BIG IS BAD

The Shrinking Man

The "Little Earthers" stand in sharp contrast to the believers in growth. Science, they say, has challenged the biblical image of the human role in the universe, moving humankind from center stage to only one species among millions on one planet of a middle-sized star in one galaxy among millions. A recent species at that. The most important product of the moon landings was not the opportunity to pick up a few rocks or even to show our technical muscle. It was—or should have been—the images from space of an Earth that is not just finite but almost endearingly small.

That world view has posed troubling questions for many Christians and has stirred the most visceral resistance among fundamentalists, leading to their demands that schools teach creationism. The traditional Christian and the scientific viewpoints lead to very different perceptions of the vastness of the Earth and of what humans are entitled to do with it.

Shifting Realities

The faith in growth tends to lump two trends. The growth curve in key areas stopped or changed into a decline in the late '70s, but we have not yet adjusted to change. In agriculture, growth has slowed (chapter 2). In forestry, we are savaging the last of the old stands in the name of "salvage logging" at the behest of lumber companies with a very short view of even their own interests. We cut down old growth, including the old snag trees that provide nesting sites, as we log the forests with increasing intensity. Robbed of their nesting places, the woodpeckers, nuthatches and chickadees decline in numbers. Then we begin to lose forests to insects such as the spruce budworm and the southern pine beetle, and wonder why. Efficiency unguided by a broader vision can be very inefficient indeed.

Growth has reversed in ocean fisheries, in U.S. petroleum production and in real wages, and the age of affluence has changed to an age of widening disparities. We were riding high in the United States. The letdown is correspondingly severe. We worry about the drug culture. In a sense, drugs offer a substitute high.

The Little Earthers ask with increasing insistency: "When should growth stop? Has it already gone too far?" They do not question the fact of past growth; they ask whether growth is not leading to its own collapse. They argue that many of our tensions—the current backlash against government in the United States, for example—result from growth. The regulations that generated the backlash were themselves the result of the rising damage humans are doing. The purpose was not to limit freedom but rather to arrest a decline that thoughtful people could see in our surroundings. Those who object to such controls tend to forget that in the simpler golden age to which they would return there weren't leaking underground storage tanks, liquid waste disposal systems injecting wastes into the groundwater, nuclear wastes, agricultural pesticides and all the other concomitant problems of modern industrial societies. We are climbing a steepening hill; the search for makeshift solutions becomes more difficult as human pressure on the Earth increases. Without a reduction in that pressure—from both population size and our consumption habits—the remedies will become increasingly complex and onerous, and decreasingly successful.

A Place to Stand
Growth involves penalties that are seldom discussed in serious books simply because they are very hard to quantify and difficult to treat within the standard academic disciplines. They deserve much more attention.

On a finite Earth with a growing population, the rich and powerful tend to sequester the best. In capitalist societies, this happens through the operation of the market. Communist leaderships made sure that it happened informally. The process is only partly countered in progressive societies by the deliberate setting aside of national parks, monuments and public land, and that activity is under constant siege, as Congress demonstrated in recent proposals to turn national parks and federal lands over to private interests.

One of my strong childhood impressions is of a simple beach house in the dunes on a coastal island in Georgia. It had a wonderful long, dark central room looking out past a porch and under the live oaks toward the sea. It belonged to a woman named Madge Bingham, who lived, as we would have said then, in "modest circumstances." She could not afford to live there now. There is only so much coast and a much larger population to compete for parts of it. The market awards the best sites to the people who can pay. Miss Bingham's cottage, the dunes and most of the live oaks have been replaced by an immense condominium.

The market was operable then and now. What has changed is the ratio of people to coastline and the intensified competition that it generates. The competition is expressed in land values. Santa Fe, where I live, is a fascinating microcosm of the process and its effects on relations between neighbors. The poor are displaced as population growth drives up the price of land. The competition for water intensifies, and the old shallow wells are the first to dry up. The mood grows blacker. The rich buy up the water rights from those who need the money. The less fortunate, who once tolerated the rich, become bitter as the rich sequester the water and use it for swimming pools and golf courses. The competition, in short, becomes much more bitter as it intensifies.

Crowding and Regulation

Everybody claims rights these days, even as the rights are constrained by increasingly onerous government regulations forced upon us to keep the environment livable.

Perhaps there should be a right to get a little drunk. I remember an old miner named Lyle Donahue, who lived in a desert canyon near the California–Nevada border. He would from time to time get rousingly drunk with his buddies at the Big Pine Tavern and then head home – 50 miles. One morning, after a light snow, I encountered his car tracks weaving on and off the road, across the desert and then back onto the road. I thought it was enormously funny. It wouldn't be funny now. In those days, there were literally no other cars on that road at night. Donahue did not endanger others. He could indeed have gone to sleep and frozen to death, but that was his affair. Now, although it is still open country, there are enough other cars to make an intoxicated driver a menace even in the desert. And given the damage that off-road vehicles are doing to the land, there would be good reason to penalize Donahue just for driving off onto the desert.

It may, I admit, be a bit extreme to defend Lyle Donahue's right to drink and drive, particularly across the desert (which is now a designated desert protection zone), but it dramatizes my point. The intensifying regulation of human activities, the multiplication of controls, the state's involvement in one's management of one's own life and property, are all driven by population. In a crowded society, we cannot afford the aberrations that were tolerable in less crowded ones.

Let me give a less bizarre example. People tend to live close to or beyond their means, so they will trade poor house construction and little insulation for a low initial cost. We all pay the price in terms of higher energy inputs with all their consequences—unless the builders have to build in better quality. Though we may chafe at building codes and restrictions, there is a social purpose. Population growth means more housing. More housing means more energy loss unless we impose regulation. As the problems worsen, the rules get tougher, and individuals lose the freedom to make their own trade-off between initial price and long-term costs.

Just contemplating Los Angeles' regulations is an argument for smallness. No outdoor fires, no grills or barbecues, complex traffic controls to try to avoid gridlock on the freeways, onerous but necessary rules about building and industry, rules on industrial emissions so tough that paint shops and dry cleaners must relocate outside the city (thus generating even more traffic and pollution).

The Human Scale

I submit that most people don't like crowding, except perhaps at parties. Opinion polls suggest that country and small-town dwellers are generally more satisfied than city dwellers. The 1985 American Housing Survey found that rural residents were happiest with where they live and had the fewest perceived problems (particularly if they had access to a small city). Big-city residents were the least satisfied, with suburbanites and small-city residents in the middle of the scale. "Studies of residential preferences over the past 40 years have consistently shown that many people who live in large cities would prefer to live in smaller cities, towns or rural areas."[217] A poll in my county of Santa Fe found that 89% of noncity residents preferred not to live in the city, whereas city residents were almost evenly split: 45% would prefer to live in the country; 42% preferred the city.

Anecdotal evidence supports such polls. What do old town Boston, Portsmouth, New Hampshire, Annapolis, Maryland, Georgetown, Washington, D.C., Alexandria, Virginia, Charleston, South Carolina, Savannah, Georgia, Santa Fe, New Mexico, San Francisco, and the two Portlands (in Maine and Oregon) have in common? All of them were built to a human, low-density scale; then they stagnated while modern skyscraper cities emerged. Now people are crowding in to enjoy the small scale and the sense of livability—and residents complain (rightly) that their cities are being "ruined."

Why, for that matter, do new arrivals generally want to "pull up the ladder"? Why is "a house of one's own" such a tenacious American dream? Because crowding is not a pleasant sport. Why has the great American trek to California reversed in the past decade? Perhaps some don't like immigrants, but I suspect most simply don't like "Megalopolis."

Compare the interaction of people in the streets of an average small town with the harried faces, suspicion and tense irritability of New York City streets. "Compassion fatigue" is mostly a phenomenon of Megalopolis.

I do not decry cities; I just decry very big ones. Perhaps I should create an honor roll of cities that successfully downsize. Pittsburgh was the epitome of the "rust belt." It lost half its population and most of its industry. It now has little unemployment and has become a better place to live—

a sometime winner, in fact, of competitions for the "best city in the United States to live in."[218]

If people like lower density and smaller scale, why not make them the objects of policy?

Crowding and Costs

Life on the intensive margin involves high capital costs. In most cities and particularly suburbs, the lack of "affordable housing" is a constant lament. We have forgotten the law of supply and demand. The jobs tend to be in the cities, and the cost of housing is driven up by land prices, which reflect the competition for space. People have to live somewhere that is accessible to work; frequently they are afraid to live in decaying cities, so they fight for a place in the narrow zone of accessible suburbia. Those who lose the competition pay the penalty of commuting for hours each day, which itself involves high direct, social and environmental costs. I have heard very few commuters praise commuting as a lifestyle, except when compared to the alternative. The costs of escaping rise as cities grow.

The cost of Megalopolis is hardly correlated with the satisfaction it produces. When did a multibillion-dollar elevated expressway ever bring you happiness? Or the equally expensive integrated sewage systems that you have bought with your tax dollars? What they provide is the avoidance of more onerous alternatives: traffic gridlock or foul rivers and polluted drinking water. Those problems are driven by population growth, and the "solutions" themselves are soon overtaken. Witness the grim humor about traffic jamming the freeways soon after they open. Megalopolis demands freeways, schools, hospitals, water and sewage systems and rapid transit systems built at increasing cost and disruption, as well as more police, courts and prisons. Increasingly onerous rules must be imposed to fight growing air pollution, but still the cities spread their air pollution and wastes into the countryside and out to sea.

These things have to be paid for by taxes. Cities that cannot pay for the improvements decay or collapse—even as the prosperous flee and tax rates go up to compensate for a declining tax base. This is a costly process in every way. GNP calculations help us to delude ourselves that we are gaining something, because the costs of growth—from parking meters to government-subsidized subway systems—become part of GNP. In a simpler age we paid for neither, but free parking—or better, a stroll to one's job in unpolluted air—is invisible in the GNP.

If cities are built in earthquake or hurricane zones, the costs and risks of urbanization are multiplied. As the 1989 San Francisco and 1995 Los Angeles earthquakes demonstrated, it is much more costly to rebuild elevated freeways than two-lane blacktops. Florida simply has too many people for a hurricane-prone zone (p. 66). In the United States, unlike in Bangladesh, the movement into such dangerous zones is voluntary, but the growth of total population drives the movement in both instances.

The costs of complexity begin to overwhelm us, but our thinking is muddied because we have not recognized or thought through the incompatibility of the two paradigms. Politicians seem sometimes to realize that growth cannot go on forever, but they still turn to growth to solve our immediate problems (chapter 7). The population issue awaits a resolution of that conflict and a transition to a less expansive view of Earth.

CHAPTER 15

Of Tigers, Ants and People

THERE IS ANOTHER CONFLICT of self-images—paradigms, again—that we seem to perceive only dimly. It stands in the way, not perhaps of limited policy changes on immigration, but of any sustained and rational national policy on growth.

INDIVIDUALISTS AND "SOCIETALISTS"

Some 70 years ago, Clarence Day wrote a brilliant little book called *This Simian World*.[219] In it, he identified a central reality in human conduct: unlike the solitary tiger or the communal ant, humans are partly individualistic and partly social animals, and the conflict between the two aspects of our character explains much of the tension in our lives, both creative and destructive.

That conflict may help to explain American behavior. The unresolved ambivalence underlies two competing images of the nature of society and the purpose of the state. For convenience, I will give names to the two viewpoints: "individualist" and "societal." The first term needs no defense. I use "societalist" in a broad sense to categorize those who emphasize humans' social nature, without necessarily implying commitment to any particular system. There is a loose and varying tendency among Republicans to identify with the first mind-set, and of Democrats to identify with the second—though the word would frighten them. It is hard to find a term involving the roots "commun-" or "socio-" that does not sound pejorative to most Americans. That perhaps is a measure of the individualistic bent of the country.

The divergence was explicit in the egalitarianism and redistribution of wealth espoused by Franklin D. Roosevelt. It is not immutable, but it does help to explain much of the rhetoric as people and politicians talk past each other.

> *The "individualist" viewpoint*: "The purpose of government is to keep the peace, provide for national defense, and provide a level playing field for individuals to reach their full potential. The accretion of wealth is a well-justified reward to energy, initiative and risk-taking. A sense of compassion may justify a certain amount of welfare for the poor, but those people do not have a claim on one's hard-won earnings. The individual is the measure of worth and the ultimate sovereign, the state should leave him alone except for the minimal management of necessary common projects."

One can hardly imagine President Harding or Coolidge or Hoover making Franklin Roosevelt's second inaugural speech in 1937: "I see one-third of a nation ill-housed, ill-clad and ill-nourished." They would not have seen it as a matter of particular importance—Christ said "Ye have the poor always with you" (Matthew 26:11). Many Republicans of his time were convinced Roosevelt was a Communist, but his message was rooted in a different view of society: that we all have a stake in that one-third of the nation and that the rich will not prosper if the poor remain excluded.

The individualist has yet to adjust to urbanism. He sees himself as a frontiersmen or cowboy. This viewpoint makes Americans particularly hard to govern. Observers since de Toqueville have commented on our boisterous national character as compared with Europeans' willingness to accept more social discipline. "Rugged individualism" resonates in the American mind much more than phrases such as "social conscience."

At the extreme individualist end of the spectrum—it would be unfair to lump them with the others—is the growing "militia movement," which prepares to take the law into its own hands and resist government intervention by force.

The nation is not going to take on population growth if it is "none of government's business." And we won't take it on—as a social issue, at least—if we don't care how our compatriots are faring.

The present mood of unrestrained capitalism has one direct effect on our population future. Business wants cheap labor. It wants strikebreakers if it is fighting the unions. And poor aliens, particularly if they are illegal and docile for fear of exposure, provide both. There are numerous and graphic studies of the uses in American history of immigrant labor to break wage scales and of the collaboration between employers and immigrant networks to supplant local hires in the meat packing industry of the Midwest, the lemon groves of California, the building trades, janitorial services, road construction and even the computer industry (chapter 11). The immigrants help their friends get jobs. There is nothing wrong with that, except what happens to the erstwhile employees.

I have pointed out that Americans, judging by multiple polls, want less immigration, but they get more. Congress goes the other way, responding to lobbying by the U.S. Chamber of Commerce, large industrial interests and agribusiness. Money talks.

> *The "societal" viewpoint*: "Humans are social animals. In a society, extremes of wealth and poverty are intolerable. The child of wealth and the child of the ghetto do not compete on a level field. Equal opportunity is a mockery when schooling, contacts, job opportunities and the ability to influence legislation all favor the rich. The rich have an obligation to give up part of their income to ensure that the poor have an opportunity and a reasonable standard of living. A nation is not just an atomistic collection of individuals pursuing their own interests. Eventually, a nation unwilling to recognize a common good cannot pursue it and becomes helpless even to protect the members' individual interests."

One corollary of this vision is the willingness, even eagerness, to mobilize government to carry out this social agenda.

Perhaps I have stereotyped the two viewpoints, but one can recognize some politicians in those "quotations."

Mutual Obligations

The Social Contract. The individualist's views are the product of some strong feelings about fairness, and they are justified because the nation has not really come to grips with the implications of the Roosevelt revolution. If society is to take more responsibility for people's well-being, those people have responsibilities—which they often do not admit—as well as the rights that they are quick to claim. The social contract must run both ways. A society cannot function without the mutual acceptance of obligations as well as rights, and one sees very little of that in either position. The individualists are perhaps entitled to be a bit madder than the "societalists," since the chances are that their taxes pay for most of the abuses.

Health insurance is a good example. As the costs rise, the irritation of those who pay the burden goes up. If we are to assume a social responsibility for people's health, people should behave in ways that do not wreck their health and increase the burden. Proponents think of health coverage without any context, as an abstraction: an absolute moral obligation to help those who need it. To get a more practical perspective, visit a hospital emergency room and watch the procession of injuries and acute seizures resulting from liquor, drugs, brawls and presumably from the tensions in our cities. A great deal of public expenditure on health involves taking care of people who will not or cannot take care of themselves or their families. There are old people on Medicaid with prosperous children. There are others whose children spend down their parents' assets and let them go on Medicaid. A senator from Illinois, with a mother on Medicaid, was elected even though she had cashed a check for the sale of her mother's property that should have gone to Medicaid.

The next question is, How far do we carry that reciprocal obligation? Is society, because it offers health care, entitled to demand that I stop smoking? stop drinking? eat less junk food? go out and exercise?

Nationally, we have not thought through that corollary. We have yet to find a balance between individual rights and social obligations. Having none, and impelled by different unspoken images of where the balance should be, we have come to frame the debate as a hostile confrontation between individual interests and social demands.

What Kind of People Are We?

I can understand the individualists' irritation. I cannot share their vision. That vision worked better in a less crowded age. The post–Civil War era was the golden age of the Republican Party. The West was unfenced, and the railroads had reached far enough to permit range cattle to be shipped to the growing markets of the East. It was the unfettered age of capitalism. There were about one–fifth as many Americans as there are now. There was land for the taking and forests to be cut down. It was a simpler world. Food came from nearby, transport was provided by wagons on dirt roads and by light–duty railroads, energy came from local woodlots or, later, nearby coal mines. The ratio of resources to people was much greater than it is now. It was a time when migrants from a crowded Europe could get land of their own, when the age–old limits to growth seemed to have been broken, and when it was indeed possible to live with minimal government. We did plenty of damage to the country's land and forests even then, but it didn't seem quite so important.

That unfettered era ended when Americans—led by Theodore Roosevelt, a liberal Republican—became concerned enough at the growing wealth and power of the rich to begin instituting anti–trust legislation and, eventually, tax laws to control their power.

As this is written, we are witnessing a highly individualistic reaction to the costs and restrictions imposed by 60 years' pursuit of social goals, from Rooseveltian income redistribution to the environmental movement. This counterrevolution, unsoftened by a sense of shared social purpose, is leading to a breakdown in our ability to manage our national affairs. The partial shutdown of the national government for weeks in 1995/96 is a frightening omen. There is a new mood: "dog eat dog." It is a dangerous state of mind for the functioning of a society. I would argue that the individualist is being supplanted by the reactionary, and all of us will suffer for it.

Dismembering the Sixteenth Amendment

The nation faces transformation into a plutocracy (chapter 11). We faced that peril once before, during the heyday of the Industrial Revolution, from the Civil War to the turn of the twentieth century. Fortunes were being made and politicians corrupted. One part of the response to the growing arrogance of the rich was the sixteenth amendment to the Constitution (proposed in 1909—by a Republican Congress—and adopted in 1913), which permitted the levy of graduated income taxes to relate the tax burden to the ability to pay. At first, it was seen simply as a revenue device, but it and the estate taxes that followed became instruments of social policy—to prevent the accumulation of wealth in a few hands. Now there is a reaction under way, with politicians in both parties proposing various versions of a flat tax that would undo the sixteenth amendment and encourage the more rapid accumulation of wealth and power.

These are dangerous and perhaps revolutionary proposals. To accelerate the present widening differentials of wealth assures the bifurcation of society into hostile classes. It makes the present decline of wages less bearable for those affected. The growing bitterness is already converting our country into a battleground, and the concentration of wealth proceeds.

Economics: The Ruthless Science

Our present infatuation with post-Keynesian economics is both a reflection of and a support for that individualistic counterrevolution. It glorifies the pursuit of efficiency and thereby of profits, and it would substitute the market mechanism for government direction in most economic decisions other than monetary policy. It presupposes "economic man" (p. 139), who is totally mobile, rather than real people who cannot simply shift anywhere, into any occupation, as the market needs change. It assumes the possibility of infinite substitution so as to avoid difficult questions as to when one is approaching physical limits. Most of all, it condones ruthlessness toward labor as a "factor of production." One can then, without qualms, substitute capital for labor to make management easier and to minimize businesses' responsibility to labor, culminating in the use of "temps." These attitudes drive the policies that are displacing U.S. labor (chapters 11 and 12).

For several years, the nation has been congratulating itself on the triumph of capitalism over socialism. The congratulations may be premature. There is no question of the crash of socialism as it was practiced in the Soviet empire, but modern capitalism, with its fixation on profit and growth, is merciless. I wonder how long it, in turn, will last. Even John Maynard Keynes, the arch-druid of modern economics, in a 1930 essay, looked forward to a day when people will "once more value ends above means and prefer the good to the useful . . . " But he went on to argue that "The time for all this is not yet. For at least another hundred years we must pretend to ourselves and to every one that fair is foul and foul is fair; for foul is useful and fair is not. Avarice and usury and precaution must be our gods for a little longer still. For only they can lead us out of the tunnel of economic necessity into daylight."[220]

Keynes is to be commended for his candor but questioned on his belief that expediency, as a way of running an economy, will lead to a society founded on better principles.

The marketplace is offered as the ideal way to manage the economic system. It is a very good way of guiding the production and distribution of goods, but it offers no help in guiding their allocation, in protecting the environment and resources that support us or in managing a social infrastructure that has grown increasingly polarized.

It would be ironic if Karl Marx, having failed to create Marxism as an organizing principle, should be proven right in predicting the collapse of capitalism because of the concentration of wealth and the widening of class differences.

Parallel Interests

If the individualists and societalists cannot come together, they may find a shared interest in a better population policy.

The confrontational stance is not necessary. The individualists should address population growth for their own safety and to preserve a degree of social tranquillity. If they look beyond the very short term, they should be strongly in favor of proposals that would turn U.S. population size around. The individualist—and the individualist in all of us—can only bemoan a world in which individualism comes into increasingly sharp conflict with the needs of an expanding society. The social engineering attendant upon crowding makes it harder and harder to preserve much room for "rugged individualists." They would have more room and less governmental interference in a less crowded country.

Business is driven, not by natural brutality, but by competition. "Others do it; we will go bankrupt if we don't." They have a way out. They could use their considerable lobbying prowess to persuade government to enforce minimum wages and labor standards on everybody, thus stopping the union busters from driving the competition. They could suggest lower immigration to lessen the temptation to cheat with imported labor, and they could advocate a trade policy that does not put us at the mercy of low-wage competition abroad (chapter 12). It would work. A loyal and contented labor force would be worth something, in contrast to the present standoff and the reliance upon high-turnover but cheap imported labor.

Regrettably, our individualists have not yet heard that message, mesmerized as they are by the abortion issue and the fight against "government meddling."

Don't expect such proposals from business. Finding a cooperative solution is not the American way right now. Meanwhile, we are witnessing a gathering battle as Congress for the first time faces diametrically opposed pressures from business and from states burdened by the costs of immigrants to their health, education, welfare and social support services.

The societalist should seek a smaller population for a different reason: Most population growth takes place among economic classes that require more services than they can pay for. Population restraint would make it easier to help those people. The social support system would not be endangered by its own spiraling costs, and a smaller population at the bottom means fewer poor and a middle class society. However, societalists are likely to be hung up on taboos (chapter 16) that make them unwilling to take on the population issue.

The chance overlap of interests may make possible some limited changes in immigration law, but it is hardly the basis for a coherent and long-term view of the national interest.

Individual interest is concentrated. The social interest is dilute.

At every turn, when one argues the disadvantages of population growth, one encounters a phalanx of those for whom the growth is immediately profitable. It is a formidable coalition and, without a new (or perhaps revived) sense of community and some consensus about our obligations to each other, the coalition will win. The common good almost by definition is achieved by restraint and accommodation, and those virtues are not much in evidence right now. Exactly where we draw the lines of the social contract is less important than the consensus itself.

IN PRAISE OF PATRIOTISM

On the old road between Albuquerque and Santa Fe, New Mexico, at a crossing called Budaghers, there is an abandoned roadhouse with a gas station, which was bypassed by the new interstate. Whoever owned it retreated in good order. Plywood was put over the windows to protect the place. On one of those sheets of plywood, now partly obscured by an ailanthus, he took a paint brush and wrote "U.S.A.#1!" in big, bold letters.

There is a lot of courage reflected in that little scene. I don't know who owns the place. I suspect it was an Hispanic. It was probably somebody who had had his share of lumps, culminating in a distant decision to cut off his roadhouse with a new highway. He was undaunted nonetheless. He believed in the system, and he believed in an abstraction called country.

In a fractioning age and society, we could use more of that spirit. In fact, we desperately need it. There is not much to rally the spirit in the cry "I'm all right, Jack"—i.e., I've got mine—that animates the era. Humans are at least in part a social species. In the absence of a larger group with which to identify, we identify ourselves with subgroups, whether they be urban street gangs, ethnic groups or economic classes.

It is less divisive to invest a sense of identity and altruism in one's country, to be proud of it and concerned about its well-being. At the local level, the parallel is civic pride. We need a global focus of loyalty to perform the same unifying role for societies that those concepts do for individuals, but it will be a long time coming. Meanwhile, we need the nation. We have a better chance of making it work than of inventing a substitute to perform services that we must have.

Patriotism is not presently a popular idea. There is a dawning recognition that we share one world and that "all men are brothers." The result is that arguments based on the national good are suspect. Driven too far, patriotism can indeed become jingoism. The two world wars resulted in part from the excesses of nationalism, and the word has become pejorative in many minds. Its savage cousin, ethnic bloodshed, is riding high in much of the world. We don't need more of that. Patriotism is, however, a different concept and a powerful emotion. It has been prematurely counted out.

My interest in patriotism is partly emotional. (On census forms, I perversely insist on writing in "American" where it asks about race, thus assuring, I imagine, that my data will be recorded with American Indians or "other.") My present purpose, however, is intellectual.

I have argued that the nation is becoming engrossed in a self-destructive pursuit of class and factional rivalry. I don't think that arguments addressed to the common good will be heard unless we come again to realize that we share a common identity. In the debate about population, the adversaries tend to talk past each other, because the concept of the national good is simply absent. Arguments as to the national interest in controlling immigration are dismissed as racism or xenophobia.

A sense of patriotism and a shared destiny require some adjustments on both sides. The Democrats (with some Republican help) inadvertently promoted the divisions within our society as they sought to achieve racial justice, starting with President Johnson's Great Society program. The discrimination was real, and I can think of no other way to propel people from those minorities into the middle class, onto corporate boards and senior governmental jobs and into a position to be heard. Nevertheless, reverse discrimination, like subsidies, creates its own vested interests. Our government created a powerful inducement for people to think of themselves as members of a particular minority. We needed to know who was there and how they were faring, and that led to other self-perpetuating divisions such as racial questions on job applications and the census questions about race (a thoroughly unscientific concept) and ethnicity.

Perhaps we should have written sunset provisions into those laws to terminate them when they had served their purpose. It is hardly rational for the child of a Black Ph.D. to have a hiring advantage over a White manual laborer's child. Reverse discrimination in the name of ending discrimination has a certain Alice in Wonderland quality. There may be hope in recent proposals—voicing the unthinkable—to gear the legislation more closely to individual wrongs. In any event, "affirmative action" becomes less and less meaningful as we become a nation of minorities and of mixtures of minorities (chapter 9).

As to the Republicans, the freshman Republicans elected to the House of Representatives in November 1994, with their demolition derby approach to government, seem intent on validating Karl Marx's class view of society. Their proposal for a "capital gains cut and indexation" is a rich man's dream. So is the "flat tax" proposal, which some Democrats have taken up. In the name of a balanced budget, the Republicans have embarked upon two thoroughly incompatible projects: wiping out the deficit and cutting taxes for the prosperous. They seek a massive income transfer from the poor to the most prosperous and a removal of social controls on the latter, as if to emphasize that America consists of two classes and to show which one is in control.

The rich get richer and the poor get bitter. The rich are doing very well indeed without a tax break (chapter 11). They are destroying the sense of community. There is no automatic rule as to what income disparity should be tolerable in a society, but we are widening the gap. Those at the bottom suffer first, but our society is being torn apart, and not only the poor will suffer.

Immediate personal self-interest often argues for immigration. Most of the arguments for stopping population growth rest upon societal considerations—the plight of the poor, the preservation of the land for our descendants—and not upon immediate individual or sectarian benefit. For each individual (unless you yourself are being displaced), the costs of massive immigration and population growth are gradually imposed, whereas the benefits are immediate: cheap labor and lower prices. Only if you believe that the social good is important and worth the sacrifice of some immediate individual benefits do the arguments even reach you.

Individual interest will prevail over the social interest unless it is confronted with a powerful spirit that motivates people to write "USA#1" on window hoardings and that treats others as fellow citizens rather than simply as tools or competitors.

The Social Contract

The argument is made, correctly, that reforms are not made just by the altruistic. The Philippines might still be a U.S. colony—to our regret—if American sugar producers seeking to get rid of tariff-free Philippine sugar had not added their weight 60 years ago to the voices of moralists preaching self-determination. We have heard some very practical proposals from parochial interests to reduce fertility and immigration. Let us encourage them.

One can make the argument that private interests would be a sufficient mass to change policy if people would look objectively at what population growth does to their own long-term interests. However, a policy based simply on the occasional and accidental coincidence of individual interests is not the way to a broad population policy. Immediate self-interest is more compelling than a distant vision, and decision making can simply hang up on a deadlock of opposing views—which is close to a description of U.S. politics right now. We would be a better and happier society if we could look beyond immediate self-interest to a larger view of the good of society. It is needed not just for population policy. It is essential if we are to come to a shared sense of the future (chapter 6) and of the ways that society must change to survive.

CHAPTER 16

Multiple Agendas and the Population Taboo

THE PROLIFERATION OF POLITICAL AGENDAS, coupled with a rising note of extremism and intolerance, tends to exclude the population issue. Principal among those agendas are militant feminism and a commitment to universalism among idealists (particularly among the young) that denies the United States' right to pursue national self-interest.

A population policy inescapably requires tough choices about immigration and fertility. Many people don't want to make those choices. They are pursuing other goals and dreams and see the population issue as a threat to their agendas. For such people—many of them moral and well-intentioned—the population issue has very nearly become a taboo.

Let me categorize those attitudes briefly and offer rebuttals to show that a population policy might help in the pursuit of some of those other objectives, rather than conflicting with them.

THE NEW TOWER OF BABEL

As tensions have grown, there has been an explosion of groups seeking social justice of one sort or another. Something ominous has taken place as the participants in the political dialogue have multiplied in recent years. The political landscape is full of lobbies working at cross purposes or competing for funding. Concern about population growth has been submerged by dozens of different voices pressing parochial agendas. Very few of them have made the connection with population growth. They do not see the ways in which population growth and intensified competition block their own agendas, and consequently they look upon the population movement as a competitor.

The agendas in themselves are sometimes worthy, but we head for disaster as the support for a population policy fragments and the role of population growth in the human future is forgotten. I have heard of at least two recent cases in which lectures on population and environment were blocked by mass protests by "social justice" advocates demanding that their issue be addressed first.

Representatives of American Indian groups assert that nonindigenous populations are the source of their problems and that "overpopulation is not an issue for indigenous peoples."[221] U.S. Women of Color (USWOC) takes the position that poverty should be attacked "by tackling social and economic imbalances, not just by pushing contraceptives. We don't want to wait until the 'unmet need' for contraceptives has been satisfied before realizing that we have utterly neglected to boost social and economic progress and failed to alleviate poverty."[222] Those people have yet to consider that a population policy might help them achieve their aims.

The Women's Global Network for Reproductive Rights demands that the Population Council stop research on antifertility or contraceptive vaccines. The president of the Planned Parenthood Federation of America (PPFA) told a press conference that "there can be no advancement in the world if the status of women is not improved."[223]

The new stridency and the polarization of our politics have made the traditional politics of compromise inoperable. It does no good to say to the true believer, "I agree that your cause is reasonable, and I will support it, but give me the same support for population policy. It will help you." The answer is likely to be, "My cause is the *only* cause!"

The Abortion Issue

The most immediate and politically loaded taboo concerns abortion. It has been promoted into a moral absolute, and it generates passions that frighten politicians and the more moderate into simply avoiding the population problem. It must be addressed.

In a sense, the abortion issue is tangential to population policy. The "right–to–lifers" treat the two as if they were the same, but there is no theoretical reason why fertility could not reach 1.5 children without abortion. The problem is that, in fact, no nation has done it. When people are making love, they may—regrettably—not be thinking about babies or the population issue. Abortions provide a way of correcting that oversight.

To oppose all abortions is to abandon the future to uncontrolled growth and to promote the birth of unwanted children who will face a bleak start in life. If right–to–lifers reflect on those consequences, they should be fervent advocates of contraception and sex education. "Just say no" is reasonable moral advice but not an adequate control on fertility.

There is some hope that technology will help to end this ideological slugfest. There are technologies on the horizon to prevent the implantation of the fertilized egg. They will give second-guessers a more humane way of correcting their oversight, if they act quickly. Perhaps the anti-abortionists can use their considerable energies to persuade young women to learn to act quickly.

Perhaps the real answer is moderation. I am not likely to convince the true believers, but legitimate values and objectives can come in conflict (see the discussion of China's problems in chapter 7). Few if any principles are so absolute that they must override all others. It is legitimate to urge that abortion be a last resort but not to deny others any right to use it.

"Women's Right to Control Their Own Bodies"

The more radical element of the feminist movement says that "nobody is going to tell women what they will do with their own bodies." To address population growth through fertility thus stirs up hornets on both sides: the doctrinaire feminist who fights any proposal to meddle with women's decisions about childbearing, and the right-to-lifer who equates family planning with abortion. An odd couple.

From the beginning, there has been a fault line within the family planning movement between those who promote family planning to give women that absolute control and those who are interested in family planning as a way to bring down fertility.

Women who may be quite willing to espouse governmental intervention in other issues such as pollution find themselves in this instance lined up with the Libertarians. There is, in fact, an unusual amount of clouded thinking on both abortion and birth control. Many strong individualists (chapter 15) oppose governmental involvement in personal affairs—except abortion. Many feminists are for such intervention—except regarding fertility. When strong feelings are involved, consistency is a victim.

Two desirable goals are in conflict: freedom of choice, and influencing the nation's demographic destiny. The assertion of women's right to decide the number of their children offers no prospect that population growth will stop, if women want more than two children. The fluctuation of fertility between the Great Depression, the baby boom and the baby bust is clear evidence that women do not automatically seek a fertility level based on societal needs. Every individual must decide which goal is more important: one's own wishes, or the societal good. It is self-delusion to try to escape the choice by pretending that they are necessarily compatible.

One way of resolving the conflict and escaping the taboo would be to agree that—while preserving the principle of voluntarism—it may be necessary to encourage women to choose to "stop at two." In other words, change the context in which the choice is made. Promote incentives and disincentives (chapter 13) that lead women to choose lower fertility. This might not win over the most extreme, but the alternative is to allow the extremists to block population policy.

The more extreme feminists, one may surmise, are not themselves likely to be driving birth rates up, but as we shall see, they effectively stop most environmental organizations from addressing the population issue.

Feminists and the Cairo Conference

The UN International Conference on Population and Development took place in Cairo in September 1994. It was the third in a series that goes back to 1974. The conference was largely dominated by American feminists attending as delegation members or representatives of women's or environmental groups. The Programme that came out of the conference threw dramatic light on their priorities when the feminist and population movements meet. The feminists co-opted the population movement (as it is now embodied in official UN and U.S. documents) for their own goals.

The Cairo Programme contains hundreds of recommendations about women's rights and other social issues but almost none about population. Funds for family planning are, in the fine print, directed to "maternal and child health." The administrative costs for "reproductive health" and AIDS prevention—two-thirds of their cost—are to be taken from family planning funds (Programme sec. 13.15). There has not been enough money for direct population programs (chapter 13), and the Programme would divert some of it into other causes. The assumptions are that, given unimpeded freedom of choice, the women of the world will choose the socially desirable fertility level, so money spent on women's advancement is the most effective way to pursue population goals.

These assumptions are simply speculation. As I have said, there is no justification for the assumption that free choice will, unguided, lead to the socially desirable level of fertility. It is myth masquerading as truth.

Nowhere does the Programme state that population growth should stop. Nowhere are growing countries urged to give a high priority to stopping (or even slowing) population growth. Governments are urged to "support the principle of voluntary choice in family planning" (Programme sec. 7.15). The Programme is negative about any stronger action, including "schemes involving incentives and disincentives" (Programme sec. 7.22).[224]

The Programme has gotten it backwards. At one point, it says "Eradication of poverty will contribute to slowing population growth and to achieving early population stabilization" (Programme sec. 3.15). The problem is that you can't get there from here. With the third world unemployment described by the ILO, this "solution" simply ducks the issue. A better bet would be to turn it around: a successful population policy would help reduce poverty.

The Programme calls for universal health care, universal primary education particularly for girls, plus jobs and leadership roles for women. It calls upon governments to

> increase the capacity and competence of city and municipal authorities to manage urban development, to safeguard the environment, to respond to the needs of all citizens, including urban squatters, for personal safety, basic infrastructure and services, to eliminate health and social problems, including problems of drugs and criminality, and problems resulting from overcrowding and disasters . . . to promote the integration of migrants from rural areas into urban areas and to develop and improve their earnings capability by facilitating their access to employment, credit, vocational training and transportation . . . [and on and on].

It calls for good governance, democracy, minority rights, more international aid, and almost every imaginable good—except slowing and stopping population growth.

All this is urged in an era when unemployment is rising, national budgets are caught in the increasing costs of dealing with pollution and unemployment and their consequences, living standards are declining, and many governments are impotent in the face of overwhelming immediate problems. You cannot spend enough to achieve those goals when you don't have enough even now. You must identify the things that will do most to match the reality with the dream.

As William Catton has remarked, "to exalt goals that unalterable circumstances make unattainable is to play with dynamite."

The Cairo conference explicitly rejected the "narrow" population programs, including those developed by the U.S. Agency for International Development, that have had some success in lowering third world fertility. Heretofore, the focus has been on assuring the availability of contraceptives and knowledge about them, promoting awareness of the population problem and rallying indigenous leaders to promote population programs.

Longtime Sierra Club population activist Judith Kunovsky has given perhaps the most telling response to the argument that population will take care of itself if we take care of women's needs and social inequities. At a Sierra Club round table, it was proposed that in order to deal with population growth "We need to relearn how we treat each other, how whites treat blacks, how men treat women, how rich treat poor, how educated treat uneducated . . . The patterns of dominating the oppressed . . . " Ms. Kunovsky responded: "The problems you are describing have been around for thousands of years, and we don't even have a hundred in which to make this stupendous change . . . Resource shortages and population-driven problems are already causing violence throughout the world as people retreat into their own ethnic-religious-national groups for some measure of security. The earth's ecological system, on which we depend for survival, is already threatened . . . We really have very little time to make very dramatic transitions."[225]

Most modern Americans endorse the principle of women's equality, and population advocates have an interest in bettering women's status. Apart from its intrinsic merit, it would probably eventually contribute to a reduction in fertility. The problem is that most countries are not within sight of achieving all the good things that were advocated in the Cairo Programme, and the resources of the richer countries cannot do it for them. If—as I believe—stopping population growth is a condition precedent for the eventual achievement of those goals, and if funds are not unlimited, the issue for a population conference should be, What uses of limited funds would be the most cost effective in lowering fertility? At Cairo, they undertook instead to subvert existing population programs and to divert resources and attention from population programs to women's issues.

"ONE WORLD"

It is not easy to be up against optimism, idealism and short memories. Those of us who warn about population growth face a reaction that is particularly difficult to counter. When talking to groups that are interested in poor countries, in the environment, in solar energy or in better living conditions for the poor—in short, precisely the idealistic people who look beyond immediate self-interest and are concerned about human welfare—I often find that my message is seen as negativism. Most of these people are young; the world they see is, to them, the way the world has always been. They do not remember when it was less crowded. They want to pursue concrete and perhaps manageable ideas: do something specific about hungry people or trees or a park or bicycle paths or housing or whatever. They do not want to worry about problems so vast and apparently unmanageable as population. They don't want to be sidetracked into such vaguely negative ideas as fewer children or fewer immigrants. The phrase "negative growth" is a turnoff.

Universalism

Many such people are imbued with the "one world" dream. To limit immigration is to break the faith. They take seriously the idea that we live on a shared planet and that we are all brothers and sisters. To them, an "alien" comes from outer space, not from Mexico. To that generation and those of their elders who are on their wavelength, any proposal to limit immigration conjures up images of nativists, racists and skinheads.

There is a critical point to be made here. These are good people, but they must realize that the population issue, like many others, is not a battle of unalloyed good against unalloyed evil. It is a choice forced upon us by the reality that resources and the environment are not infinite. It is, to use the cowboy movie image, a battle between "white hats." To put it bluntly, the choice is between an obligation to the stranger and the obligation to our own disenfranchised and to the environment our children inherit.

Though the choice is between desirable objectives, it is far from symmetrical, a point usually ignored. We cannot save the world, but we have a fighting chance of dealing successfully with our own problems.

A letter to the editors of *Sierra* pretty well summarizes the universalist state of mind. The writer argues that "What victimizes black workers today is not the influx of immigrants, . . . but the flight of industry to other countries where labor is cheaper. . . . [I] do not suggest support for endless growth in the United States. . . . [but] to treat the symptoms by closing national borders to keep out 'the disease' rather than commit ourselves to eliminating the cause of the illness, is both futile and irresponsible."[226]

Those brief sentences are driven more by passion than analysis. They call for an exegesis.

1. Both immigration and the departure of capital hurt the American labor competing with immigrant or cheaper foreign labor. Change has been drying up jobs in American industry (chapter 11). The letter writer is arguing that increasing the supply of a factor of production (labor) in a situation of glut does not lower the price of the labor. He thus denies the principle of supply and demand. Polemically, of course, the denial makes it possible to blame bad guys

(industrialists) and to absolve oneself of any obligation to limit immigration or worry about American labor.[227]

2. Projections cited in chapter 9 make it clear that to be for unlimited immigration is in fact to espouse "endless growth in the United States."

3. There is a certain unconscious arrogance in the assumption that the United States can eliminate "the cause of the illness." The illness, presumably, is poverty and population growth. The United States should do what it can to help the afflicted countries. To assume that we can somehow save them requires a remarkably inflated view of our power. Later in this chapter, I will present some numbers that make it clear why we alone cannot eliminate "the cause of the illness."

4. The writer offers a false either/or choice. In fact, we *can* simultaneously limit immigration and provide support to family planning and self-help efforts in other countries. We will be better able to help others if we are not overwhelmed by our own labor, urban and environmental problems. To refuse to address the problems at home in the belief that we should instead solve those in the rest of the world is a very questionable bit of reasoning. One might remind the letter writer of another rule of thumb beloved of environmentalists: "Think globally; act locally."

Innumeracy

(Garrett Hardin's memorable neologism.) How often I have heard people resist the idea of a population policy because "it's not just a matter of numbers," when they don't know the numbers. In fact, demography is supremely a matter of numbers, and those numbers affect the pursuit of most personal and social goals. Comparative national math proficiency tests repeatedly demonstrate that Americans by and large are less able to deal with numerical concepts than other nations' students. The deficiency makes it difficult to reach people with numerical arguments.

The writer to *Sierra* is hardly alone in the thought that "I don't want to focus on numbers and quotas but to solve the problem at the sources." In 1991, in answering the question "Is immigration the major cause of U.S. population problems?" a National Audubon Society/Population Crisis Committee booklet evaded the question but said that pressures to migrate to the United States can be relieved by "investments in economic development and family planning in migrant sending countries . . . "[228]

Well enough, but they did not calculate how much that might cost or suggest what we should do for the next 20 years; most of those migrants are already born. This is more of a crutch than a policy proposal; nobody looks seriously at the question, Is it really possible?

The third world working-age population is rising more than 60 million per year (figure 8-1) and will continue to do so for the indefinite future. For Central America and the Caribbean alone, the annual increase is 2.7 million, nearly twice the comparable figure for the United States. They must find the resources to absorb more new working-age people than we tried (less than successfully) to do as the baby boomers came on the job market—and they must do it in economies that, in total, are one-seventeenth as large as ours. A rough calculation suggests that a total investment approaching $200 billion each year would be needed to provide the additional employment in that one region; even this does not provide for any progress in employing the presently unemployed, and the result of the effort would be to face the same problems in more crowded societies the next year.[229] That investment is more than half their entire GNP. They can't afford it. How much of that capital could we provide, and what happens to our own unemployed as the capital goes abroad to employ foreigners?

The population problem must be addressed where it exists. We should do what we can for others, both for their own good and our long-term benefit. However, that is no answer to our present immigration problem.

Relativism

Any proposal to address population growth encounters the response, "but you must take cultural differences into account." True, but that applies to the way in which the population argument is presented and to the approach that different cultures might take to limiting fertility. It does not somehow vitiate the impact of population growth on a society. (Interestingly, this "cultural difference" argument disappears when advocates are pressing their own agendas. At the Cairo population conference, the feminist agenda was hardly muted in deference to Vatican and conservative Arab views.)

Alienation

Other attitudes play a role. Many people ferociously criticize their own society and idealize others. I am hesitant to get into psycho-history, but there does seem to be an element there of hatred and rejection of one's own society. I attended a discussion meeting recently at a Unitarian church. The speakers were so angry at the U.S. Government's Central America policy (as they understood it) that they demanded we bring the "victims" here. This seems a remarkable case of cutting off one's nose to spite one's face. The Americans who would suffer from such a policy are not the particular political leaders who have angered the speakers, but rather our country's dispossessed and our descendants.

People who are solicitous of the foreigner are strangely impervious to the problems of Americans. It may be easier to sympathize with the stranger, who is an abstraction, than with our own poor, who are a sometimes troublesome presence.

The Vatican

David Simcox (himself a Catholic layman) has written a fascinating short article describing the increasingly uncompromising position the Roman Catholic Church hierarchy has taken on the right to immigrate and on the obligation of the United States to accept unlimited immigration. The American Bishops lobbied for the Immigration Act of 1990, which raised immigration. Pope John Paul II in 1987 even endorsed the "sanctuary movement," which undertook to hide illegal immigrants in violation of U.S. law. Church writers assert that "Catholic citizens are required to work to see that as far as possible the laws of their countries adhere to this universal norm" of the right to migrate.[230] Coupled with the papal opposition to limiting family size (p. 209), this puts the Vatican squarely in opposition to any effort to limit U.S. population growth.

From the church's standpoint, this universalist and pronatalist viewpoint may seem justified, but it strikes at the heart of the social contract discussed in chapter 15 and it makes it very difficult to pursue a rational population policy. Unfortunately, church guidance on temporal matters is most likely to be taken seriously among those who traditionally are most fertile.

DENIAL

Denial is the crosscutting argument that unites those who want to avoid the population issue for whatever reason. If you have a problem but don't want to face it, deny it exists. If you find it painful to reduce the emission of greenhouse gases, simply tell yourself that the climate warming theory is still just a theory, and you don't have to worry about it. Deny there is a population problem, and you do not face the dilemmas it poses.

"Jobs Americans Won't Take"

A common example of denial is the argument that immigration is needed to fill menial jobs.

The rebuttal to that one is pretty straightforward: it is not the jobs; it's the conditions. In 1993, the nation watched the arrival of ships full of

Chinese whom even pro-immigration Senator Kennedy called "slave labor." This is the labor that apologists say fills the "jobs that Americans won't take." Does anybody of goodwill really believe that Americans should work for as little as 70 cents an hour and be locked up every night in a crowded and windowless room? As with other commerce, the lowest bid tends to set the price of labor, and this indentured labor drives down wages and takes jobs from other unskilled people in the cities. Of young Americans not in school, only 54% have regular full-time jobs (figure 11-1). We should begin to think of our own people.

Sugar cane cutting is, I believe, the only category of labor in the United States in which Americans do not constitute a majority. "Jobs Americans won't take" is a euphemism for "wages Americans won't take." Pay a living wage, and we can employ our own unemployed (p. 186).

Those who are being hurt have made their choice, despite the taboos that afflict the ideologues. The poor and the urban minorities do not have the luxury of distance. Among those most affected—but who might be expected to welcome more immigration for reasons of ethnic solidarity—are the U.S. Hispanics, most of whom are for less immigration (p. 261).

"The Government Can't Do Anything Right"

This attitude is a form of alienation. It would deny a social role in population policy on the grounds government would do it wrong. Sometimes it simply means "government costs me too much" or "doesn't serve my interests as I see them." There is, nevertheless, growing resentment as government "meddles" more in an increasingly crowded society.

My response is, Who would you have looking after social issues, or making tax policy, or watching out for the common good? Business? Nobody? There are some roles that must be performed by society at large, and that function is embodied in government. If it doesn't work right, try to fix it. Anarchy doesn't have a very good record either.

Immigrant Contributions

The enthusiasm for liberal immigration comes from an odd alliance: special interests, and an academic climate in which immigration is politically correct. Since they see it as racist to limit immigration, apologists need to reassure others that it is good for us. The nation has been treated for years to the argument that "immigrants put in more taxes than they take out in welfare," despite official data such as I cited on page 194. Several scholars and think tanks (notably the Urban Institute) have for years tried to prove that immigrants do not displace unskilled U.S. labor. The proposition is counterintuitive and flies in the face of studies of various occupations discussed earlier.

The argument about tax contributions is especially silly because it is irrelevant. Taxes do not simply pay for welfare. They pay for environmental protection, roads, the administration of justice, national defense, public land management, education, public libraries, police protection, public hospitals, street cleaning and all the other services that government provides. Taxpayers who do not cover their share of those burdens, aside from any welfare they may draw, are not carrying their part of the load. Nobody has argued that recent immigrants carry that burden, but this is hardly a reason to blame them. The poor, by and large, do not pay their "share" of taxes, and most immigrants are poor; that is why they had to leave home.

The governors of California, Florida, New York and Texas, who must cover the bills, have pretty well ended the argument that current immigrants are on balance net contributors to state and local government. (The governors are demanding federal help, since it is the federal government that is responsible for controlling immigration.) But the same academics now argue that immigrants may be a net cost to the states, but on balance they contribute to the federal budget. One tends to "prove" one's preconceptions.

The argument is not against immigrants; it is about numbers. Immigrants (including the ancestors of all of us) have made great contributions to America. Uncontrolled immigration exacerbates the unemployment and social problems we presently face. The two statements are compatible.

"Taking Care of the Old"

The argument is regularly heard that we need more children or more immigration to take care of the growing numbers of the old. This is the equivalent of a pyramid club or Ponzi scheme, or like having a little drink to cure a hangover. The "aged boom" ahead is the product of the baby boom, and only time will cure it. Look at that "bulge" in the U.S. population pyramid in figure 7–1b. That is the baby boom. You cannot cure a baby boom with another boom; that simply defers the solution until the population problem is much worse. Indeed, as I suggested in chapter 12, the dependency ratio in this country will rise only briefly because of the baby boom and then settle at a reasonable level.

This is, on reflection, an odd problem to be worked up about when our problem is the displacement of labor by capital and technology, a decline of wages and idle people wandering the streets and getting into trouble.

Price as the Measure

Pro-growth economists argue that things are getting better because the price of some commodity—usually food—has been going down for decades, as measured by the number of hours the average employee must work to earn a given quantity. This has been true of some prices (but by no means all) in the industrial world, reflecting the secular rise in productivity and—until recently—the rise in wages. For food, prices seem to be turning around now, as the supply (chapter 8) lags behind effective demand (i.e., demand by people who can afford to buy).

Perhaps a more dramatic answer to that argument is that in the Great Depression, prices were very low—for those who still had jobs. A lot of people were hungry because they had no work. One has to ask, Prices are lower for whom? And in what currencies? As we have seen, in Africa, per capita food consumption has gone down even as prices were subsidized to keep them low; much of the population is simply out of the system. Worldwide, prices are rising for those things that are under particular resource restraint—e.g., lumber and fish. When energy prices start up again (chapter 3), the general price effects will be more pervasive than during the oil scares of the '70s.

The price argument is of course irrelevant to most of the problems that lead me to argue for a population turnaround.

"Overconsumption"

The overconsumption argument (chapters 2 and 5) is a form of denial, but in this case a legitimate goal is substituted for the effort to solve the population issue, rather than complementing it. As we have seen, conservation and technical fixes offer considerable room for savings in energy but much less in agriculture. It may sound foolish for a population advocate to tell a conservation advocate that his program is politically unrealistic; the U.S. population seems thoroughly resistant to both arguments right now. I will argue (chapter 18) that the two should be working together, not in opposition.

As a subcase, it is argued that world food shortages are not a result of overpopulation but of maldistribution. For the answer to that one, go back to chapter 2.

"RACISM" AND RACISM

The most well-meaning people are terribly anxious to avoid any appearance of denigrating other races. That's fine. But it leads to some remarkable taboos. The editor of an environmental journal for which I wrote an article asked that I avoid the word "fertility" because it might be taken as veiled criticism of Blacks, Hispanics or American Indians, even though there was no reference to them in the article. He knew that they are more fertile than non-Hispanic Whites, and he knew that they know. Thus, to avoid offense, one must avoid a concept central to demography. This is like covering one's ears to avoid hearing naughty words. It doesn't contribute to a rational approach to population growth. I hope that the discussion of fertility by race (figure 13–2) is sufficient reassurance that fertility is not a racial issue.

The editors of the January 29, 1996, *National Review* remarked that "a racist is someone winning an argument with an immigration enthusiast." I rather like that.

There is real racism. It should be fading, but I don't think it is. The United States, along with most of the world, seems to be in a blind valley between belief systems right now, and the most elemental loyalties supplant the more reasoned ones. Racism leads to competitive fertility. Some Whites see the nation as being lost to others. Some Blacks resent both majority Whites and the Hispanics who are supplanting them as the largest minority. Hispanic groups with colorful titles like "La Raza" (The Race) have not forgotten the Mexican War and may see Hispanic population growth as a vindication and restoration of what is rightfully theirs.

For a chilling look at where we may be heading, look at the passions in Quebec. The provincial premier blamed "money and the ethnic vote" for defeating the recent proposal for independence. Another leader decried "the declining birth rate for Quebec's white population." A professor complained that Quebec's French women "used to have 10 to 12 children; now they have one."[231] I hardly dare project Quebec's population with a TFR of 12. Fortunately, the professor does not dictate Quebec's fertility rates.

Some groups engaged in ethnic wars—Croats, Serbs, Somalis, Afghans, Hutus and Tutsis—may with some justification think they need high fertility to produce cannon fodder, but it is a pity to see tribal hostilities override the most pressing need of our time.

People with any of these mind–sets may be willing to endorse proposals to limit the fertility of others or limit immigration of other groups. That is a bad reason for the policy, and it builds hostilities for the future.

■ ■ ■

We face a formidable combination of taboos. Taken together, they lead to a vast silence about population policy.

Chapter 17

The Failure of Leadership

THE UNITED STATES NEEDS LEADERSHIP if we are to neutralize or convince the advocates of growth and do something about population. The two best candidates to provide it are environmental groups and our political leaders. Some leaders in both categories see the population problem, but almost none of them have dared to take on the forces that drive it.

THE TIMID CRUSADE

The big environmental organizations, which should be leading the crusade to bring U.S. population growth to a halt, see the dangers and say they believe population growth should stop. During the 1988 presidential campaign, a coalition of 18 environmental organizations prepared a "Blueprint for the Environment" for the guidance of the incoming U.S. administration. The document included the statement that "U.S. population pressures threaten the environment all across our nation" and gave some examples. It said that family planning and the availability of contraceptives must be expanded worldwide. It recommended "an official population policy for the United States" and said that "We must assure that federal policies and programs promote a balance between population, resources, and environmental quality."[232]

So far, so good, but they do not say how that goal might be pursued. I repeat: the only two variables available to influence population growth are fertility and migration. That proposition seems self-evident, but it poses an apparently insuperable stumbling block to the major environmental organizations. None of them has been willing to endorse specific policies on immigration or fertility, and only one—The Wilderness Society—has recently called for lower immigration and fertility.[233]

Why is it so difficult for environmentalists to gird up? I can offer only speculation. The population question is fundamentally important. Generalizations without specific proposals are an invitation to more growth. The taboo against discussing specific measures must be broken. I hope I do the organizations no injustice in the following description of their concerns. There are answers, if the organizations choose to take on the issues.

Immigration

We take seriously our self-image as "a nation of immigrants." Universalism is probably strongest among idealists, who are the soul of the environmental movement, and the environmental organizations are afraid of antagonizing their support. The debate has been plunged into a region of moral absolutes, where to compromise is to sin. This is a powerful deterrent to organizations that rely on the enthusiasm of the self-righteous.

The *Earth Island Journal* carried a "declaration" that population and immigration are the "least" of the causes of environmental damage, that illegal immigration is the fault of the industrialized countries, that immigrants help rather than harm the economy, and so on.[234] Faced with such ferocious advocacy and concerned about retaining their own constituency, the mainline environmental groups address population only in anodyne generalities and avoid positions on immigration.

Environmental groups are quite capable of resorting to *denial* and *innumeracy* in order to avoid taking a real position on immigration. I have cited one example (p. 244). Immigration is a matter of numbers. To take refuge in cliches about solving the problem at its source—coupled with silence when immigration legislation is on the Congressional docket—is to accept the consequences. Environmental groups are not shy about talking numbers where other environmental concerns are at stake; the timidity about immigration results from ambivalence.

The largest environmental groups and foundations also have a practical problem. They look toward family planning cooperation with counterparts in other countries to save the shared planet, and they fear that supporting immigration limitations may threaten that cooperation by raising the suspicion that they are acting in narrow self-interest.[235]

The fear is misplaced. Other countries (including Mexico) do not propose that their borders be opened. They recognize their limits, if not ours.

Fertility

For the largest organizations, with substantial conservative memberships, the biggest obstacle in addressing fertility is probably the visceral fear of being drawn into the abortion issue. Most of the big organizations, including the National Wildlife Federation and the National Audubon Society, take no position on abortion. So far as I know, there are no survey data to guide them in deciding whether anti-abortionists belong to environmental organizations in any numbers. I rather suspect they do not. This is one of those issues that the organizations will have to face. Is it more important to keep the marginal members or to address the fertility issue?

I have pointed out that there is no necessary connection between abortion and fertility levels. For the more timid environmentalist groups, I would offer a compromise that is better than no position at all: advocate more and better family planning and the financing to make sure it reaches the poor. To those members who oppose abortion, they can make the point that family planning should actually help to reduce abortions.

The dread charge of "racism" is another issue to which environmentalists are particularly sensitive. The environmental movement in fact is very close to being "lily white," and it is trying to reach out to people of color. Figure 13-2 provides the answer: high fertility is a function more of poverty and lack of education than of race. Whatever reduces those evils helps everybody and particularly minorities.

The environmentalists have no problem with family planning abroad. What is sauce for the goose should be sauce for the gander. The U.S. Agency for International Development (AID) for years has had a program ("Plato" and its successors) to show third world leaders why they must bring fertility down, plus a program to show them how to organize to do it. Those programs are right in line with the Blueprint for the Environment (p. 251), but the U.S. environmental movement does not offer comparable leadership abroad, and it provides none at home.

Divided Feminists

Environmental groups face a problem within their own ranks from the advocates of women's absolute right to control their own bodies (chapter 16). Those advocates constitute a considerable element within the environmental movement. Environmental leaders don't want to antagonize them, and their influence hinders the adoption of population platforms. I have argued that "friendly persuasion" might make a good compromise (p. 240), but compromise is not popular right now.

The organizations are caught in a monumental non sequitur. The attempt to straddle the feminist divisions and the silence on immigration do not lead to the population results they endorse. The major environmental organizations may or may not reflect the concerns of their constituency, but they cannot claim that they are doing much about population growth.

Alliance Building

One group's public affairs officer defended its caution in staking out stronger positions on the grounds that the first priority is to enlist other groups—women's organizations, labor, ethnic and minority groups—in the population cause. She said those groups are suspicious of a strong position on immigration or fertility.

The concern is understandable, but her organization may in the process lose its own purpose. A priori, it would seem better advised not to weaken the message but to show the other groups how an effective population policy serves their interests. With labor or minority organizations, for instance, the "hot button" is likely to be unemployment and urban problems. The same is true of Black organizations. With some 61% of young Blacks neither employed full time nor in school, those organizations may find it hard to identify with environmentalists who talk about trees and wilderness and endangered species those young people have never seen. Talk instead about the kids' plight and you might find new allies. I have seen nothing to suggest that the big environmental groups make that connection.

The Refuge of Avoidance

The environmental groups become particularly susceptible to denial and to "solutions" that aren't enough. Conservation is a particularly popular theme in environmental literature. Environmental groups put out pamphlets with titles such as "Twenty Ways to Save the Environment" that, quite literally, tell you to turn off the water tap while you brush your teeth (I am not joking) but say nothing about considering the number of children you plan to have or writing your Congressional delegation about pending immigration legislation. There is no harm in marginal suggestions about toothbrushing if they are not used as a way to escape talking about population. If, however, the environmentalists content themselves with such trivialities and avoid a role in immigration policy or fertility where they might have some impact, we will all lose.

THE POLITICS OF CAUTION

Our government was far more forthright in addressing the issue of population growth a generation ago than it is now. Meanwhile, we have grown by 60 million people.

The Heroic Age

In 1969, President Nixon raised the issue of U.S. population growth. Let me quote him at some length, even though readers may have seen this quotation before. His was a prophetic vision from, perhaps, a surprising source.

> In 1917 the total number of Americans passed 100 million, after three full centuries of steady growth. In 1967—just half a century later—the 200 million mark was passed. If the present rate of growth continues, the third hundred million persons will be added in roughly a thirty-year period. This means that by the year 2000, or shortly thereafter, there will be more than 300 million Americans.
>
> The growth will produce serious challenges for our society. I believe that many of our present social problems may be related to the fact that we have had only fifty years in which to accommodate the second hundred million Americans. . . .
>
> Where, for example, will the next hundred million Americans live? . . .
>
> Other questions confront us. How, for example, will we house the next hundred million Americans? . . .
>
> How will we educate and employ such a large number of people? Will our transportation systems move them about as quickly and economically as necessary? How will we provide adequate health care when our population reaches 300 million? Will our political structures have to be reordered, too, when our society grows to such proportions? . . .
>
> . . . we should establish as a national goal the provision of adequate family planning services within the next five years to all those who want them but cannot afford them.[236]

The current President might well be urged to repeat that final recommendation as the first plank in any program to address welfare reform and the exploding problem of pregnancy among unmarried teenage girls. Here we have Richard Nixon and Joycelyn Elders (p. 201) hand in hand. Quite a surprise.

President Nixon proposed and Congress later created a distinguished National Commission on Population Growth and the American Future, chaired by John D. Rockefeller III. In 1972, the Commission concluded that the country would benefit if population growth stopped. It made many explicit recommendations. Among them:

- that the government establish a permanent long-term strategic planning capability to monitor demographic, resource and environmental trends, serve as a "lobby for the future" and recommend policies to deal with looming problems.

- that preparations be made to deal with the anticipated growth of metropolitan areas by reforming governmental arrangements it characterized as "archaic."

- that there be a national, publicly funded "voluntary program to reduce unwanted fertility, to improve the outcome of pregnancy, and to improve the health of children." This was to include education, contraceptive information and—delicate as the subject was even then—abortion.

- that immigration levels not be increased, that those levels be periodically reviewed to see if they are too high, that there be civil and criminal penalties for employers of illegal aliens, and that the enforcement program be strengthened.[237]

The Rockefeller Commission recommendations were quickly swept under the rug. They were controversial and it was an election year. Let us not forget that blunt, honest population study. The recommendations would be a good starting place even now for organizations seeking to make realistic proposals as to how to address the United States' future.

Eight years later, the *Global 2000 Report* to President Carter described some of the connections between U.S. population growth and resource and environmental problems. It offered no recommendations, but it was followed up by a booklet of action proposals from the U.S. Department of State and the Council on Environmental Quality, which included eight broad recommendations concerning U.S. population growth.

The United States should develop a national population policy which addresses the issues of:

- population stabilization

- Availability of family planning programs

- Rural and urban migration issues

- Public education on population concerns

- Just, consistent, and workable immigration laws

- The role of the private sector—nonprofit, academic and business

- Improved information needs and capacity to analyze impacts of population growth within the United States

- Institutional arrangements to ensure continued federal attention to domestic population issues.[238]

These proposals were less specific than those of 1972, but they reiterated the same themes.

We have been slipping backward since then. Business wants cheap labor. Would-be ethnic leaders want to enlarge their constituency. They play upon others' guilt and the mystique of "a nation of immigrants," so immigration policy is paralyzed. Politicians are intimidated by the ferocity of right-to-lifers who confuse family planning with abortion and who have promoted their view to an absolute, demanding the right to impose it on others. Immobilized on these two issues, Washington is unable to address population change. It is a particularly dangerous abdication of the government's oft-abdicated responsibility to serve the people.

From Hostility to Ambivalence

The Reagan administration was openly hostile to the population movement. It tried at first to discontinue all foreign population assistance, and when Congress resisted, the funding was provided by continuing resolution. It suspended assistance to the UN Fund for Population (UNFPA). At the UN Mexico City conference on population, the U.S. representative astonished world leaders by declaring that population growth was "neutral" in development. Only the Vatican representative agreed.

The Bush administration insofar as possible simply avoided the issue as too hot to handle, though Bush in earlier years had been openly supportive of world efforts to stop population growth. The Clinton administration has supported increased family planning assistance to poor countries (p. 213), but it has avoided the issue of U.S. population growth.

Even those politicians who abstractly recognize the population problem instinctively revert to growth as a solution to the problems facing them. President Clinton has said some of the right words about population growth. On Earth Day 1993, the President observed that there may be 9 billion people on the planet in the future and that "its capacity to support and sustain our lives will be very much diminished."

When faced with current problems such as unemployment, however, he reverts to the thoroughly American and Romantic idea of growing out of them. His March 14, 1994, address to the G-7 was particularly revealing of his mind-set: "There is no rich country on Earth that can expand its own job base and its incomes unless there is global economic growth. In the absence of that growth, poorer countries doing the same thing we do for wages our people can't live on will chip away at our position. When there is a lot of growth, you can be developing new technologies, new activities and new markets. That is our only option."

President Clinton gave a speech on population on June 29, 1994, to the National Academy of Sciences. He emphasized the need to invest in women. He added that "reducing population growth without providing economic opportunities won't work" and that the population problem would be pursued "as part of the larger issue of sustainable development." He said that "Our population policy is rooted in the idea that the family should be at the center of all our objectives," but he did not elaborate how that relates to population growth. In short, to be unkind: bromides.

The speech reflected the President's internal dichotomy, a mind-set that has not yet absorbed the lessons of a finite Earth under environmental onslaught. He would solve every problem by encouraging growth. "We're going to talk about what we can do within the G–7 to promote not just growth, but more jobs—because a lot of the wealthy countries are finding they can't create jobs even when they grow their economy."

On world population, he said variously that "you must reduce the rate of population growth" or (ambiguously) "stabilize population growth." Few people in the population movement would consider those to be sufficient goals. If the *rate* were halved, third world population would still double every three generations. The Earth could not sustain it.

Somebody should tell the President that perpetual growth is impossible on a finite planet. We must find other solutions.

The Vice President understands third world population problems. In his impassioned and moving book *Earth in the Balance* (1992), he said that "No goal is more crucial to healing the global environment than stabilizing human population." But he was thinking of the third world. As to the United States, he simply listed it among the countries "with relatively stable populations." He has not considered recent projections of U.S. population growth. Nor has he made the connection between immigration and population growth, at least in public. Underlying his book is the perhaps unconscious assumption that third world population growth will simply stay in the third world. If Vice President Gore is still at that stage, one may be pretty sure that U.S. population is a nonissue for most of Washington.

The PCSD

There is one small exception. On June 14, 1993, the President's Council on Sustainable Development was created by presidential order. Its mandate was equivocal from the start. The President charged it with helping to "grow the economy and preserve the environment . . . ," objectives that are likely to conflict. It is a mixed body including cabinet members, environmentalists, labor leaders and industrialists of different sexes, races and ethnicities. Population and consumption—two of the critical elements of sustainability—were initially not even in its scope. They were introduced, over opposition, at the instance of Council member and Under Secretary of State Timothy Wirth.

To its credit, the Council, in its report presented in March 1996, called for a "Move toward stabilization of U.S. population" (Goal 8) and said that "The United States should have policies and programs that contribute to stabilizing global human population . . . " (Principle 12). It remained equivocal, however, as to the broader issue of growth itself. " . . . some things must grow—jobs, productivity, wages, capital and savings, profits . . . " It did not address the likelihood that more workers with higher productivity will stress the environment even more, even with efforts at amelioration, and that growth itself is at some point unsustainable. It did not, in other words, explore what sustainability really is. It avoided the problem of how to stabilize population, by leaving fertility decisions to "responsible" individual decisions and avoiding positions on abortion or on immigration levels.

The President was "pleased . . . to accept" the report and asked the Council to stay in being through 1996 to work on putting the proposals into effect under the leadership of the Vice President. This is a small footnote to history. The report and the President's acceptance of it, casual as it was, constitute the first explicit acceptance by a U.S. President of the proposition that U.S. population should stop growing. Let it be said in their behalf that the Council members showed that even such a diverse group can recognize the need to stop population growth. They are not alone in their inability to face the tough decisions that would be needed to do it.[239]

Such commissions are regularly used to defer action on tough issues—they are "being studied"—and the Council was no exception. Ironically, it submitted its report just as Congress was considering whether to lower immigration levels—and the White House, as we will see, was undercutting the effort to reduce immigration. So much for the advice even of one's own Council.

The administration has not really connected population growth and immigration policy, and there simply has been no policy on U.S. population and the fertility of American women.

Cairo and the U.S. Government

The United States' role at the ICPD at Cairo in 1994 reflected the Clinton administration's priorities. Faced with the militant feminists, the administration decided to join them rather than fight them. Department of State Under Secretary Tim Wirth remarked of the United States' role that "women's groups pretty much drove this."[240] The result was outlined in chapter 16; the Cairo Programme has much more to do with feminist goals than population policy.

If the administration uses the Cairo Programme for guidance, U.S. policy on third world population growth will consist of rhetoric more than practical help, and the hard-won funds for family planning assistance will be fair game for any organization whose agenda can be construed as "women's and reproductive health" or even the more general Cairo objectives such as helping minorities, children, the aged, or indigenous peoples or "eliminating poverty" (see p. 241).

Unfortunately, the U.S. Government played an active role in creating the Programme, now endorses it, and has modified its statements accordingly. Tim Wirth probably understands the dangers of population growth better than most U.S. political leaders, but he sometimes shifts his rhetoric to accommodate his new constituency. In a March 30, 1994, speech on the Cairo Conference, he said that "the empowerment, employment and involvement of women must be the overriding catalyst for common purpose in . . . Cairo. . . . At the end of this century, the extent to which we fostered the transition to sustainable development will be measured in part by our success in refocusing scarce resources and redirecting national priorities on behalf of women . . . " He listed seven objectives, all of which were directed toward women's issues; none touched on population growth, and only one of them included a reference to "voluntary family planning." A good speech on women's rights but not on population.

The U.S. delegation agreed with the Programme's language on migration, perhaps the most politically explosive demographic issue in the world today. After a brief skirmish, a third world proposal to declare a "right" to family reunification across borders was somewhat altered to call upon receiving countries to "recognize the vital importance of family reunification" and adjust their laws accordingly (Programme sec.10.12). The point is far from minor. U.S. immigration policy since 1965 has centered on the reunification of extended families in the United States. The right of immigrants, once they are citizens, to bring in their siblings and *their* families has been the central cause of the subsequent growth of immigration. Any effort to control immigration must address family "chain migration." The government's acquiescence in the ICPD formula mirrors its lack of interest in the subject.

The Republican Alternative

None of the present national Republican leaders has shown any interest in population as such. Domestically, they are saddled with the abortion issue, which leads to an unwillingness to take on the population problem.

As to population aid to others, there is much opposition on that side of the aisle to all foreign aid, and an early casualty has been U.S. foreign assistance in family planning (see chapter 13).

On the other hand, Republicans have taken the initiative on several issues that affect U.S. population.

Population Legislation

I have described some Republicans' interest in discouraging pregnancy among women on relief, which has the inadvertent demographic side effect of promoting lower fertility (chapter 12).

As to immigration, recent Congressional history has been one of division on both sides of the aisle. Immigration reform bills have been introduced in most Congresses since the '70s, when we discovered the consequences of the ill-starred immigration act of 1965. There is no doubt as to where public opinion stands or of Congress' willingness to ignore it. Congress passed the Immigration Act of 1990, raising immigration by 40% or more, in the face of public opinion polls that regularly showed a large (60% to 80+%) majority of Americans, including Hispanic ethnic groups, calling for less immigration.[241]

In 1996, Congress scuttled a bipartisan effort led by Republicans to reduce immigration. The relevant bills are H.R. 2202, introduced by Congressman Lamar Smith of Texas and S.R. 1394, introduced by Senator Alan Simpson of Wyoming. The bills in their original form would have significantly reduced immigration. They were gutted over their sponsors' objections. First, legal and illegal immigration were divided into separate bills in response to the immigration lobby. Then the proposal for a better identification system was watered down to a limited "pilot program," which in the House version was explicitly voluntary. This change will put off the critically needed system of identification (p. 211) into some uncertain future. All that was left was authorization to increase the Border Patrol— and authorizations do not necessarily result in appropriations—plus minor reforms such as making financial guarantors legally responsible for immigrants' support.

As of this writing, reduction of legal immigration has been deferred indefinitely, and the slight differences between the Senate and House bills on illegal immigration still have to be resolved. Some form of legislation will probably emerge before the elections to try to fool the public into thinking Congress is "doing something" about illegal immigration.

The scuttling of the original immigration bills took place just after a new Roper poll indicated that 83% of the respondents were in favor of reduced immigration and that 70% favored a level less than 300,000 (which is less than one-third the current level). Only 2% wanted more immigration.[242] The hearings showed why the popular will gets thwarted. Congress was subjected to a major assault led by business groups. The country's major business organizations came out against changing the law on legal immigration, including the U.S. Chamber of Commerce, the National Association of Manufacturers, the National Foreign Trade Council, Associated General Contractors, CitiBank and several major export-oriented corporations. They formed an organization called American Business for Legal Immigration, and its members fanned out through Congress to influence the decision makers. Senator Spencer Abraham, (R-Mich), produced a list of organizations opposing reform that featured "business groups and high-technology organizations that oppose limits on employers' abilities to bring in foreign workers."[243] Even the measures ostensibly intended to protect American technicians from displacement by low-wage foreign workers were more sound than substance. Senator Simpson had some meaningful measures in his original legislation but was forced to drop them during committee hearings. The Specter/Abraham amendment was substituted, which was carefully worded to be useless in operation though noble in expression.[244]

The administration equivocated before and during the legislative process. Attorney General Reno responded to rising criticism by making some administrative changes. The INS budget was increased by 72% from FY1993 to FY1996, largely for better control of illegal immigration, but Reno has said that "the administration is committed to maintaining the U.S. tradition of liberal immigration policies . . . we will not permit this cherished tradition to be jeopardized by weakness in enforcing our immigration laws."[245] It is curious that she is so committed to a "tradition" that has existed fitfully, is partly fanciful (see chapter 13) and in recent years has flouted the popular will.

There have been clear signs of internal conflict within the Clinton administration about immigration policy and the question of identification. Testifying before the Senate Subcommittee on Immigration and Refugee Affairs in September 1995, Immigration Commissioner Doris Meissner said the administration wanted a new immigration bill before the year's end and favored a ceiling lowered to 490,000. That proposal sank like a stone, with neither Meissner nor any other administration figure repeating it. Later, to please immigration advocates, she and the Attorney General urged that proposals for reduced legal immigration be removed from the two immigration bills.[246] The administration was conspicuously silent as the proposal for better identification was watered down to nullity.

One is entitled to doubt whether the U.S. Government, as responsive as it is to powerful interests, will seriously address population growth in the foreseeable future.

Passivity

Passivity is a broader and more general problem that paralyzes any effort to avert population growth. Politicians and planners, like scientists (p. 72), regularly treat population as an "independent variable" that must be accommodated but cannot be changed.

This attitude can be seen at all levels of government. The planners accept a projection of population growth for the next few years and then calculate how many roads, schools and sewage lines are needed. They do not ask where that leads us. What will our prospect look like then? Will we simply have to build more of everything, forever? They do not explore how that population growth might be avoided. This mind-set endures, even where the voters have elected city and county officials on "no-growth" tickets. When I chided a planner for that perspective, the retort was "planners lose their jobs when they begin to talk like that."

It does indeed require a change of thinking habits, particularly for a local government to see how it can influence population growth, but the change can be made. Any county, for instance, is entitled to develop zoning that would offer the hope of keeping population and water supplies in balance. Beyond that, county representatives should tell the state's elected representatives in Washington about the effect of national policies such as immigration laws on their own situation. They would, in a wiser world, also ask what fertility in their own county is doing to growth.

So far as I know, the only argument that interested the states and counties in the recent Congressional debate over the immigration bills was the potential effect on their welfare and public health costs. They do not see the underlying population argument: immigration drives up population and creates an endless need for more services and more growth.

CHAPTER 18

Creating a Consensus

WHAT IMAGES RESIDE after this long journey? To summarize:

- First, the thoroughness with which population and economic growth penetrate modern political, social and environmental issues.
- Second, the wide variability of the resultant impacts, geographically and on different issues.
- Third, the dangers and complexities of allowing the Earth to run as a single economic system, contrasted with . . .
- Fourth, the necessity to see the Earth as a unified ecosystem and humankind as a part of that system rather than its master.
- Fifth, the impossibility of growth as a long-term solution, and the multiple benefits of reversing growth itself rather than trying to accommodate it.
- Sixth, the relative simplicity of that solution, contrasted with . . .
- Seventh, the immense resistances to undertaking it.

In this final chapter, let me propose some ideas as to how we in the United States could learn to take on the issues of growth, especially population growth.

In chapters 14 through 17 I suggested the process will not be easy. Even if the popular mind can be won over, there is a question whether popular opinion alone can change national policy in our emerging plutocracy. Special interest groups, the rich and the powerful must be convinced, too, if we are to find a consensus. The poor already know they have problems, but the stock market soars even as the less fortunate are in trouble, and that makes it hard for stockholders to focus on more distant problems.

How do we climb this mountain? I don't think anybody knows, but I think the crux is to persuade those who are presently very comfortable that the growing problems threaten them, too. With that in mind, let me offer some thoughts.

MAKE GOVERNMENT RELEVANT

Foreign policy professionals, friends from earlier days, periodically complain to me that support for international programs is threatened and that "we must get people interested in foreign policy." The answer, I believe, is that government had better make foreign policy relevant to the public. People are worried about jobs, and plenty of them understand that NAFTA and the ITO—major policy objectives in Washington—are no friends of theirs (chapter 12); environmentalists share those feelings. Instead of addressing such concerns, foreign policy activity is frequently an effort to catch up with mistakes or the pursuit of matters irrelevant to most of America. I will give an example of each.

1. Ad hoc decisions made in the White House generated the surge of *Haitian and Cuban boat people* that set out for the United States in 1994. President Bush in May 1992 had ordered the repatriation of Haitian boat people intercepted at sea to forestall their landing and claiming asylum. Presidential candidate Clinton denounced the policy. In April 1994, stung by civil rights groups' criticism of his failure to change the policy once in office, he allowed several hundred Haitians ashore for interviewing. The word got back to Cuba and Haiti, and the 1994 boatlift began. Panicking, the White House on May 8 announced that claimants would not be allowed ashore but would be interviewed aboard ship or at third country "safe havens." This just encouraged the boatlift since it offered the opportunity to be picked up and interviewed without a dangerous ocean trip. Reversing course again, the White House in July canceled the May 8 announcement, but the damage had been done. The U.S. Government spent the summer stuffing Cuban and Haitian boat people into our Guantanamo naval base and pleading with nearby countries to accept some of them.

   We wound up that messy series of improvisations by promising the Cuban government to take 20,000 immigrants a year—a unique commitment without sanction in our law. We moved the occupants of the Guantanamo camps to the United States—after having declared we would not do so—and closed the camps in February 1996. As to Haiti, the affair forced us into the subsequent military action to overthrow the military junta and restore President Duvalier in order to unload the boat people—actions we had carefully avoided until the boat people forced our hand.

If we had had a clear image of what our immigration policy should be and had developed contingency policies (see below), we could have avoided the entire exercise. As to the bombast about "returning democracy to Haiti," it is one of the most desperate nations on Earth, stripped of its forests and soils, with a total fertility rate of 4.8 and an infant mortality rate of 8.6%. It is a tough place to start learning about democracy. The Haitian exodus will most likely resume, presumably leading to our return to President Bush's policy. It would have been a good place to start helping with a population policy—about a generation and 3 million Haitians ago.

2. *Bosnia.* International intervention in Bosnia was occasioned by the best of humanitarian motives but informed by a deficient sense of history. The Balkans have been the Balkans for a long time. One may indeed sympathize with the terrible suffering there, but Americans clearly wonder whether military intervention there is in our vital interest—any more than, say, in Rwanda and Burundi, where we draw back from intervention. (There is a NATO rationale for our Bosnian involvement, but it is very much an inside-the-beltway argument.) If there is one vitally important aspect of the Bosnian tragedy, it is that we and the Russians (who tend to like Serbs) avoid being drawn into a confrontation with each other and that Moslem support for the Bosnians not imperil the oil trade. One world war started in Sarajevo. Perhaps we need to move back to a less ambitious and quieter agenda of localizing eruptions such as that in Bosnia, rather than trying to impose a cure on those who don't want to be cured. Wasn't one Somalia enough?

A time of reflection might allow us to identify the really vital issues affecting America, such as the environment, employment, trade and immigration. In speech after speech, Secretary of Labor Reich has been saying many of the things I have said in this book, but nobody listens because secretaries of labor are supposed to keep organized labor quiet, not make policy. Perhaps he should be invited to discuss his concerns in the National Security Council.

Some reflection would also give us time to think ahead. Notice that I did not include Desert Storm in my criticism of foreign policy. It was vital. If Saddam had moved into the military vacuum of Saudi Arabia, the whole industrial world would have been in deep trouble. But having escaped one crisis, we need to find ways to reduce that vulnerability, and that is what much of Part I of this book was about.

CREATE A FORESIGHT PROCESS

Perhaps we can train ourselves to face the issues. Society has a short attention span, and people are not likely to focus on the slow but powerful processes of demography. I am not optimistic that human nature will change much. We humans tend to be driven by mysterious and contradictory feelings and prejudices. We are chronically unwilling to consider the likely results of what we are doing. However, we are quite capable of making rational choices when we force ourselves to look at the consequences. There are ways to do that.

Moving a Reluctant Government

I have described three areas where U.S. decision making simply avoids looking at the consequences of demography: Middle Eastern policy (chapter 8), trade and unemployment (chapter 12), and Haitian migrants (above). There are strong political and bureaucratic pressures not to widen the focus. How do we bring demography into the national thinking as the government goes about making decisions? There is no groundswell of support for a population policy. There is no reason to hope that politicians will get out in front of public opinion to lead this cause, or to expect that they will automatically think of population when they are addressing seemingly unrelated policy decisions.

Politicians after President Nixon have not been ready to take on population and growth issues frontally. Perhaps they can more safely approach the issues by asking, in one context after another, What are the impacts if we don't stop?

We need mechanisms to bring the practical consequences of growth into focus on a case-by-case basis—a conscious national process of self-education, if you will. The process would improve governmental decision making. At the same time, if the decision process can be made more open so as to reach the public, it would help to move population up the scale of people's awareness. Once people have seriously thought of the connections between immigration and population growth, they will probably never look at immigration in quite the same way again.

Foresight

There is a cure for tunnel vision. The process of systematically looking at the potential consequences of what we are doing is called foresight.[247] In the Introduction to this book, I cited the colossal problems generated by the failure to address third world fertility when mortality started downward. Only by identifying and dramatizing the consequences of such failures can we overcome the fears that lock the nation.

The foresight process is not limited to population issues, but these represent the area in which it may be most needed. Population policy is regularly made by policy makers who are thinking about other issues.

Immigration Policy as an Example. In the Cuban and Haitian boatlift, systematic foresight machinery might have led to some contingency planning and forewarned the President of the likely result of his April 1994 announcement.

Immigration policy, when it has reached the White House at all, has been dominated by the usual "big players": the National Security Council (NSC) and the Departments of State, Defense, Justice and Treasury. Many fundamental consequences are not addressed when decisions are made on issues thought of as solely "foreign policy" or "Justice Department" or "security" issues.

Presuming on an ancient NSC connection, I urged the National Security Council, without success, to take the lead in creating a permanent interagency policy group on population and immigration. It could give other departments and agencies a voice in the decisions. Immigration impinges on almost every area of government, both directly and by its effect on U.S. population growth. Almost every department is (or should be) interested in immigration policy. For example:

- The Department of Labor needs to share its perspective about the impacts on jobs and wages.
- The Department of Agriculture should be heard on the connections between U.S. population growth and food supplies and exports.
- The Housing and Urban Development Department should be asked how the additional population, much of it at economic levels that will need subsidized housing, will affect current housing availability and future housing budgets.
- The Health and Human Services Department could bring several interesting perspectives into the debate: tuberculosis and AIDS control; the effects on Social Security; public health costs; and local, state and national welfare budgets.

Such an interagency group on migration could inform the decision process when issues arise such as the boat people. It might even influence the thinking when Congress undertakes one of its periodic "reforms" to increase immigration in behalf of one or another interest group.

The White House held some ad hoc interagency consultations after the boat people crisis had started. The machinery should be made inclusive and permanent, at a policy level, and charged with addressing long-term issues.

There are antecedents for such a policy group. In the '80s, following the creation of the Domestic and Economic Policy Councils, interagency groups were created to address crosscutting issues such as acid precipitation (p. 59). In the Carter administration, there was an NSC–chartered Interagency Task Force on Population and a less formal Interagency Committee on International Environmental Affairs, both chaired by the Department of State and both (unfortunately) dealing only with international aspects of population and environmental policy.

The creation of permanent working groups on specific issues would begin the foresight process. There are ways the process could be made more systematic.

The Environment Impact Statement (EIS). There is one legal instrument already in place to address population growth. It is the National Environmental Policy Act of 1969 (NEPA). It begins with an eloquent statement of the principles I have been trying to convey in this book.

> The purposes of this Act are: To declare a national policy which will encourage a productive and enjoyable harmony between man and his environment; to promote efforts which will prevent or eliminate damage to the environment and biosphere and stimulate the health and welfare of man; to enrich the understanding of the ecological systems and natural resources important to the Nation . . .
>
> The Congress, recognizing the profound impact of man's activity on the . . . natural environment, particularly the profound influences of population growth, high density urbanization, industrial expansion, resource exploitation, and new and expanding technological advances . . . declares that it is the continuing policy of the Federal Government . . . to create and maintain conditions under which man and nature can exist in productive harmony, and fulfill the social, economic and other requirements of present and future generations of Americans.

The act then goes on to spell out a process, which came to be known as the EIS process, to examine proposed governmental actions to assure that they meet the purposes of the act.

The act did not tell the government what it could or could not do. It was a "process bill." It told the government that it must consider the potential environmental consequences of proposed actions. And it enlisted the public.

NEPA did not permit exceptions. It applied to all governmental actions. Unfortunately, from the start, the government was more diligent about applying it to routine projects such as highway construction than to major decisions, and even that diligence has been waning. The EISs were considered an inconvenience and were written to justify actions rather than to explore their consequences. Despite the clear language of the act about population, judicial interpretations have been mixed as to whether the demographic consequences of a proposed action must be considered.[248]

NEPA contains all that is required to begin the process of self-education as to the role of population growth in the pursuit of national policies. It could be applied by any president or cabinet officer who chose to employ it. It could be required by private suit. It is, after all, the law of the land.

The Critical Trends Assessment Act. Following the failure of the government to use the NEPA process, other proposals for foresight machinery were made.

Vice President Gore during his time as congressman and senator repeatedly introduced a bill called the Critical Trends Assessment Act. The proposed act was the best of several proposals to create foresight machinery in government to address crosscutting issues such as population growth. He submitted the bill because the *Global 2000 Report to the President* in 1980 made it clear that the nation lacks the capability to cure its tunnel vision and see the side effects of what it is doing.

The heart of the proposal was to create a process in the White House to ensure that new and unfolding national and international trends are identified and their policy implications assessed. It was not a perfect bill. For one thing, it did not contain a provision for the automatic scanning of proposals reaching the White House to consider their side effects. Nevertheless, it was a good start.

Sponsorship by the White House of a bill to create a broader and better thinking process might just catch the public eye. It would certainly fit in with the Vice President's work on "reinventing government." He pushed for it for years until he became Vice President and was muzzled by discretion. It might even be endorsed by the Republicans. Several years ago it was cosponsored in the House by Newt Gingrich, of all people. This might be one of the few bills that could get top-level bipartisan sponsorship in the present political standoff in Washington.

That suggestion is fueled more by hope than by expectation. Nevertheless, a break in the standoff may be possible. It would be an act of statesmanship either to go back and make use of the EIS process or, failing that, to create new machinery to make us hold an ongoing debate on the effects of population growth. It would be a good job for a vice president.[249]

The foresight concept is not just my personal enthusiasm. It has had widespread support in the past. It is the only specific policy proposal in which most of the environmental and population community is formally united—in a resolution more than a decade old.[250]

A New EPA Proposal. The Science Advisory Board of EPA in January 1995 issued a set of recommendations titled *Beyond the Horizon: Using Foresight to Protect the Environmental Future.* It started with the recommendations that

> As much attention should be given to avoiding future environmental problems as to controlling current ones. . . . EPA should establish an early-warning system to identify potential future environmental risks.

First among the "forces of change" they listed was this:

> The continuing growth in human populations, and the concentration of growing populations in large urban areas.

I wholeheartedly agree.

OPTIMIZE; DON'T MAXIMIZE

All sorts of intangibles are usually left out of the population equation: the beauty of an unpolluted sky, the feel and smell of a clean wind.

Should I admit to a fondness for raw oysters on the half shell? It is becoming increasingly dangerous to indulge that taste. Oyster beds are being wiped out by pollution, despite heroic efforts to control it. The remaining oysters are dangerous to eat, particularly raw oysters (unless you enjoy hepatitis). This may sound like a sybarite's lament, but it does make a point: maximum is not optimum. We may be able to support the present population, with its sewage running into the nation's wetlands and bays, and warn people not to eat oysters or fish from polluted rivers. There are other things to eat. But is this gradual process of deprivation a goal? or simply a necessity?

Remarkably, the population debate is almost always conducted in terms of how many people the Earth, or a given part of it, can support. Is that the objective? Maximum population leaves no room for error or for a period of poor harvests. Even sustainability, a popular idea these days, makes no room for the question, Sustainable at what level?

One regularly hears U.S. consumption levels compared with those in places like India as if the United States were doing something wrong. Having spent three years in India, I do not believe that 936 million Indians living at their consumption level are better off than—let us say—150 million Americans living as we do.

I have argued (chapter 14) that crowding imposes social controls that are not needed in less crowded societies. The Earth with a population of 2 billion could sustain a lifestyle that only the rich can sustain as the population passes 6 billion. Those 2 billion people could maintain their lifestyle without insulting the environment or angering the less fortunate.

As the noose tightens, it becomes "immoral" to do things that once were benign.

Golf courses are a case in point. On a sparsely settled Earth, there should be no particular objection to them other than their managers' enthusiasm for pesticides and herbicides. In the real world, they divert land that others need and become a focal point for envy. From a casual reading of news stories, I have collected reports of protests against proposed golf courses in the American West (because they compete for scarce water), Mexico, the Philippines, Taiwan and Malaysia (because they take peasants' land). There is no more obvious symbol of disparity of wealth and the competition for limited resources than a golf course.

I have discussed the increasingly complex and onerous proposals to save water in the West. As population presses ever more relentlessly on resources, the conscientious propose desperate efforts to achieve greater efficiencies. The efficiencies we pursue may turn sour. In New Mexico, there is about one-tenth the water per person that there was at the turn of the century, simply because of population growth. Then, they could afford the wonderful luxury of cottonwoods along unlined irrigation ditches, and birds in the cottonwoods. The ditches are being lined or converted to pipes—or they will be, in the relentless search for water for the cities. The cottonwoods go and along with them the birds that used to help control crop pests. So we substitute chemicals and find them turning up as carcinogens in our water supply. Is this a transformation that intelligent beings would choose?

We fail to step back and ask, Why are we being driven?

An optimum population will not live in luxury—not for a long time, at least. Even if the nation recognized the benefits of a smaller population, it would be generations before we achieved it and could say "we have protected the biosphere; now let's live it up."

Rather, we should be seeking the reasonable, asking ourselves, What is a decent standard of consumption for all, and how many people can enjoy it, after we have assured that we are not degrading or monopolizing the environment? In short, What would be the optimum population for the United States? There is no scientific process for determining optimum population. There are too many uncertainties and too many value judgments. This is a question of philosophy and judgment, which science can inform but not decide.

As a nation, we have not yet begun the inquiry. Several years ago, I posed the question to specialists in different fields: "From the standpoint of your area of interest, what would be the optimum population of the United States?" From the bottom of the low estimate to the top of the high estimate, the range was 40 to 250 million. I concluded that 150 million is a reasonable target. That is where we were at the close of World War II.251

The precise number is not important. It would be hard to imagine the nation reaching a consensus on any specific figure. The point is to note the direction: we agreed that we would be better off with a population smaller than the present one. This is a positive formulation of the conclusion that the Rockefeller Commission reached in 1972. Armed with that consensus, the nation could begin to formulate a population policy and then test our consensus as we proceed.

Perhaps we could turn the question around, as I did in chapter 6: What arguments are there for *more* people? At the social level, I can think of none other than those discussed (and I hope effectively rebutted) in chapter 14.

BUILD A MODEL

Critics of current economics (myself included) have not offered a concrete, practical alternative to the perpetual growth model. The addiction to growth is so deep that its devotees should be shown a model, including the Keynesian monetary flows, of a system that functions without a growing population or physical growth. One iconoclastic economist, Herman Daly, has coined a name for it: the "steady state economy." Nobody has yet sketched out such a model, so far as I can learn. Japan, Europe, Singapore, Taiwan and Korea are on the way to demonstrating how to do it, but their leaders are distinctly nervous at the experiment.

A key issue will be that of jobs. Growth is embraced to create jobs in the face of the multiple forces generating unemployment and lowering wages (chapter 11). A legitimate retort, suggested in several earlier chapters, would be that it is harder to create enough jobs *with* population growth. There is, however, a legitimate issue here: Can a stable economy create enough jobs for even a stable population, given the growth of technology and the diminishing need for workers?

The idealized system must

- provide the investment to ameliorate environmental pollution;
- generate the productivity to support the aged and incapacitated, with enough jobs to provide work for a work force that would eventually stabilize;
- function on a sustainable basis, protecting the agricultural resource base, water supplies, the ocean and the atmosphere; and
- assure some balance between the operation of the profit motive and the protection of environmental and labor standards.

Perhaps if there were more thinking and analysis on these lines, those who fear stagnation would see that there are alternatives to growth until we crash.

LISTEN TO THE BIOLOGISTS

It is not hard to understand why laymen turn to economists for expert opinion on population. In a society that worships GNP, economists are the high priests.

We should send John Maynard Keynes back to monetary policy. We can no longer afford the viewpoint that sees the Earth as an infinite source of raw materials and a dumping place for wastes from economic activities. We need to learn from those who deal with the Earth—biologists and those in related disciplines—not with limited subsets like monetary and fiscal policy. The conventional economists are out of their depth when they undertake to deal with the world external to their own financial and monetary loop.

Keynesian economics leads us into a false view of the human condition. It assumes perpetual expansion in a fixed environment. Herman Daly's steady state economics recognizes that perpetual growth is impossible in a finite environment. He has illustrated the point with a graph that brilliantly reminds us that the economy is a subsystem, not the entire system.

As Dr. Daly has explained, the point is that "To put it starkly, in the neoclassical view the economy contains the ecosystem; in the view advocated here (call it ecological economics), the ecosystem constrains the economy to which it supplies a throughput of matter–energy . . . according to some rule of sustainable yield rather than according to individual

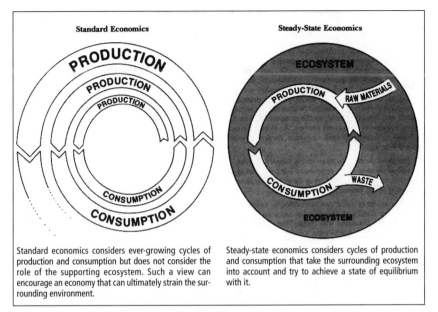

Standard Economics	**Steady-State Economics**
Standard economics considers ever-growing cycles of production and consumption but does not consider the role of the supporting ecosystem. Such a view can encourage an economy that can ultimately strain the surrounding environment.	Steady-state economics considers cycles of production and consumption that take the surrounding ecosystem into account and try to achieve a state of equilibrium with it.

Fig. 18–1. A Revised View of the Economy
(*Courtesy Herman E. Daly*)

willingness to pay."[252] He tells an amusing story of the peremptory rejection by a senior World Bank economist of his suggestion that the traditional graph (left side) in a World Bank report be enclosed with a box to show that the economy functions within a larger environment.

Daly points also to conventional economics' flawed faith in compound interest, quoting Frederick Soddy—a rare example of a scientist (chemist) turned economist: "You cannot permanently pit an absurd human convention, . . . compound interest, against the natural law of the spontaneous decrement of wealth (entropy)." Over time, unless physical production keeps up with the expansion of money resulting from interest, you face inflation or default. Like the argument about importing labor to support the aged (p. 248), it is a "pyramid club"—borrowing larger and larger amounts of resources to keep solvent—and it requires an unjustifiable faith in perpetual growth. "The market," says Daly, "by itself has no criterion by which to limit its scale relative to its environment."

Keynesian economics, when extrapolated beyond its monetary limits, is something of an oxymoron. It postulates a stable system (the circular flow of money) in pursuit of instability (perpetual growth). When we deal with the Earth, we need the opposite conceptual approach: recognize the instability that accompanies growth, and seek stability—"sustainability"—in Earth's life support systems.

The biologists are the experts on what human growth is doing to the Earth's systems and to our own future. They have a long honor roll, starting perhaps with Aldo Leopold and Rachel Carson. Listen to them.

A "grand old man" of demography, Nathan Keyfitz, has offered a novel thought as to how to reconcile the debate between biologists and economists:

> I have to stress our ignorance of just what we are doing to the terrestrial support systems on which we press; . . . we may be preparing a sudden disaster. . . . it is fortunate that some actions can be clearly specified . . . that unambiguously make the human population more secure and prosperous, irrespective of what the ultimate total solution to the overall environment problem may turn out to be. Of these, reducing the population by having fewer births than deaths is the easiest and surest.
>
> Economists have insisted that no one yet knows enough about those limitations for economic policy to take the environmental constraints into account . . .
>
> Indeed uncertainties abound on these and other matters. Yet few seem to have noticed that every economic question from whether or not to buy a dozen eggs . . . to national policy on money supply has to cope with uncertainties. It is not explained why uncertainties have become such an obstacle to serious thought and action just on one matter of supreme importance—the environment. And on what is central to that, the role of population in damaging it . . . .
>
> The device I propose for getting the debate off dead center is to ask each side to accept, just for the sake of argument, the facts and conclusions presented by the other side. To ask, that is, IF the biologists' worst horror stories are true, what WOULD BE the right economic policies? And IF the economists' characterization of the merits of markets and the incompetence of governments are correct, what action would biologists recommend to avert catastrophe? . . . Economists would accept the facts and estimates made by biologists, who on their side would give up the hope that governments would act against the wishes of their constituents, however noble the purpose . . .[253]

An excellent idea. Ask the biologists to help us define the problem, and invite the economists to offer ideas as to what to do about it.

ENLIST CONSERVATIONISTS

I have been skirmishing in this book against those who would avoid the population issue by simply calling for conservation and a less wasteful society. If, however, conservation is seen as an essential part of a more complex reality, it is an important tool.

My point in chapter 6 was that we need a much broader view of conservation than is generally expressed. We need to look at the wastefulness of whole economic processes and choose among them accordingly, and that means that conservation must be informed by foresight (p. 267). We need to reform our pricing system to reflect externalities and make those processes pay their true social and environmental costs. More fundamentally, we must integrate conservation with a sense of optimum population so as to avoid the absurdity of setting poverty as a goal (p. 271).

One can in theory justify almost any level of individual consumption if it is counterbalanced by a sufficient population reduction. Less consumption would not necessarily be a virtue in a different world. I once argued that "The big automobile becomes an environmental threat only when the number of automobiles exceeds the capacity of the atmosphere to buffer the exhaust. The steak is an immoral diversion of resources only when population outruns the capacity of the land to support a beef diet." The only condition is that such a lifestyle demands that total population be very small.[254]

We have, instead, had to make a virtue of necessity. The moralistic argue that the lifestyle to which most other people aspire is sinful because it cannot be pursued by everybody without irreparable harm. And they are right, right now. Population is unlikely to recede in the foreseeable future to a level that would justify those big cars and steaks. We are so far from that point that we will need to conserve during the lifetime of anybody reading this book.

Conservation is a necessary companion to international policy. We can hardly continue to waste as we do while we urge population policies on the third world. We in the United States would be wise to pursue conservation as an offering to the rest of the world and evidence of good faith from Earth's most conspicuous consumers, as we urge them to accept the need for some sacrifice to preserve our shared environment (chapter 8).

Conservation is justified for its own sake. There is no justification for waste that can be avoided without impairing living standards in a world where many are poor. This principle applies to the rich and the poor. In Indian villages that have acquired piped water, I have watched the villagers leave the tap open and the water running across the bare earth, and I hoped that the water would find its way back into the aquifer.

Conservation may be as complicated as reorganizing the transportation system (p. 94) or as simple as turning off a water tap. If it can be sold as a way of saving the public money, it will have broader appeal than as an abstraction. To make that case, utilities must be required to see that water, electricity and fuel are metered at the consumer level. Energy conservation becomes much more appealing if it is proposed in the form of cheap and reliable public transport, particularly in poorer areas.

Conservation, however, is a diminishing resource. At some point, if we rely on reducing per capita consumption rather than combining it with a population policy, we will find ourselves in a more straitened world, asking, Where do we go from here?

Conservation may be the one proposal that is harder to sell than a population policy. Most people who advocate it have never tried to live on the minimum wage. There are a lot of voters out there for whom conservation would be the last of all objectives. They will not limit consumption voluntarily. Conservation will probably happen, but by duress—falling incomes reduce consumption.

Still there are ways that a consensus can be built. A liberal may be interested in the environment, while a conservative is interested in costs. Waste is costly to the taxpayers because of the cost of cleanup. It makes business less competitive. Overpopulation is expensive.

Conservation, like a population policy, is good for the environment. The one thing I urge is this: do not be beguiled into thinking it is a substitute.

RECRUIT THE KNOWLEDGEABLE

We face a circular problem. Creating a consensus requires the development of a shared vision, which requires education. In the present political climate, however, it is unlikely that population policy will find its way into many curricula or into national debates. Immigration proponents and those with other agendas suppress such debate. Fertility is taboo. The population movement itself has been diverted from the subject of population to the parallel one of women's rights. The process of education is stalled.

Others, elsewhere, have been willing to take on population as an issue. I have cited the public statements of third world leaders. Those in the United States who see the problem may face a difficult choice between candor and politics. I cannot believe, for instance, that Albert Gore is as unconcerned about U.S. population growth as his public demeanor suggests. His silence suggests he recognizes the political danger of the issue. In Congress, we are losing some of those who recognize the U.S. population problem and say so, such as Congressman Anthony Beilenson of California. Of those on the national scene as of this writing, only Under

Secretary of State Timothy Wirth addresses it regularly and publicly—and within the limits I described in chapter 17.

I cannot make the choice for those who know the problem but do not speak out. In the absence of effective political leadership and of the involvement of advocacy groups now silenced by timidity (chapter 17), books such as this only provide a rationale for an educational process that may not happen.

A Somber Optimism

THE CHINESE TAOIST CANON starts with the words *dao ke dao fei chang dao*

"The road that leads on is seldom taken."

We must deal with growth before we are overwhelmed by its consequences. Controlling our demographic destiny, rather than having it determined for us, is the most important task on Earth.

Is there really any hope that humankind can find its way through the complexities I have described? Will we come to recognize the critical role of population growth in blocking the road ahead? Can humans, attuned to immediate issues and seeking quick solutions, come to perceive the importance of a movement so majestic and—in our time frame—slow as demographic change? Will we come to understand the effects of our voracious species on a fairly small planet, which—momentarily, perhaps—we dominate but cannot steer? Will we be able to sacrifice present habits, beliefs and dreams of rising consumption to deal with a dimly perceived future?

THE WORLD

If there is any hope, it is only that. It is not a calculation. There are some grounds for optimism. Fertility falls with modernization. Not necessarily immediately, but in time. It has been true in the "old" industrial countries. It is happening with the "new" modernizing countries. South Korea, Taiwan, Hong Kong and Singapore have fertility rates lower than the United States'. As women's status and opportunities expand, they tend to pursue goals other than childbearing.

Those opportunities themselves may be wiped out by population growth, but here again there is some room for optimism. Some countries, through leadership and the use of incentives and disincentives, have come at least part way toward lower fertility without being transformed by modernization (chapter 1). Perhaps others will do so as they come to realize that arresting population growth is the least costly way of addressing specific problems such as food needs, climate change, the energy transition and unemployment.

Realizing all this, would a prudent person bet on the sagacity of *Homo sapiens*? At different times and in different civilizations, economic success has led to human proliferation, and then to turmoil. Will we as humans finally recognize the connection and act on it?

It is pretty certain that for most of the world's people, the future will not consist of industrialization and modernization as we know them. More will be poor and burdened by deteriorating environments and climate change. Mortality will rise. It is rising in parts of the world. Other third world countries will escape that immediate threat but will evolve into a form of industrial society marked by crowded megacities and pollution. We will all be affected.

Any student of history must have the gnawing sense that most of it has happened by accident, but that is the road to fatalism and inaction. The Juggernaut can be turned. The vision of a better life can still be pursued, but it would take a remarkable level of self-denial in the rich and emerging countries and an emergency effort in the poorest and most fertile countries even more drastic and painful than the Chinese effort, and that seems unlikely as a matter of political will and organization. In other words, expect a time of troubles.

THE UNITED STATES

Our country is still among the least crowded parts of Earth and the best endowed with land and energy resources. Our heritage, and the explosion of growth that we have enjoyed, have generated a deep-seated belief that growth is the natural order of things, but we are hardly immune from the world's problems. We must escape the fallacy that growth is our *deus ex machina*. We need a fundamental change of vision, a recognition that we are part of Earth. We need to temper the pursuit of immediate self-interest with a sense of community at different levels: national, worldwide, and indeed with the planetary biosphere.

This requires a national willingness to look at the future, to face our own population problem and to do what we can to help other societies escape theirs. They may not succeed, but success is relative. Whatever they can save of their environment will help them in the long haul, and that is important to us. We share the Earth.

We will need to toughen up. Have we grown so soft that we cannot confront tough issues ahead? There is an inescapable problem of discipline. Could we now, as the Pharaoh did, take Moses' advice and forego the pursuit of current gratification to assure the long-term good? Probably not. Not this time. Not in time to avoid much pain.

Without a conscious policy, where are we likely to wind up demographically? Growing steadily, without an end in sight. Sooner or later, the end of growth will be forced on us as it is being forced, much more immediately, on Africa. Given the resistances, our population growth is more likely to be stopped by the Four Horsemen than by conscious policy, but this big country has more time than most, and perhaps we will make the shift of vision in time. The effort is for the long term and deeply worth doing; if we fail badly enough, there is the specter of wiping out the "microorganisms . . . on which the entire biosphere depends" (chapter 4).

THE FOLLOWING WAVE

In the language of the Bible and of Teddy Roosevelt, we stand at Armageddon. But the history of humanity is a series of Armageddons won and lost. There will probably be others.

T. E. Lawrence, in *Revolt in the Desert*, wrote this metaphor of the Arab revolt during World War I:

> One . . . wave (and not the least) I raised and rolled before the breath of an idea, till it reached its crest, and toppled over and fell at Damascus. The wash of that wave, thrown back by the resistance of vested things, will provide the matter of the following wave, when in fullness of time the sea shall be raised once more.

We face a struggle against the resistance of vested things far more vast than the Arab revolt.

There is some comfort in the hope that our descendants will have learned something when the next wave is raised, even though the lesson may not be learned in time to avert an age of troubles. There have been collapses before, and flowerings following them.

Change has been particularly rapid of late, and we have not managed to keep up and to harness the Juggernaut of population imbalance and its twin, a consumerist technological civilization that has lost touch with the Earth. The result has been a growth of turbulence toward an unknowable end. Perhaps, even if we do not escape the gathering tragedy, out of it, with clearer vision, humankind will fashion a role on Earth that will bring us back into balance with the rest of the planet.

In the rise and fall of human affairs, and despite some black pages in the history of this century, an optimist is entitled to hope there is a trend toward a better condition and that people can learn from their mistakes. As an act of faith, let us believe that—while the human experience may be a helix—the axis of the helix is tilted upward. That hope is worth holding and working for.

The Thai used to set candles adrift at night in tiny craft on the swift-flowing Menam Chao Phraya. I don't know whether they still do, in the tumult of modern Bangkok, but it is a lovely memory. I set a hope afloat on a vaster, swifter and more troubled stream.

Here, in no particular order, are the official and semi–official compilations of data on which this book are largely based:

- United Nations (UN) Population Division, *World Population Prospects: the 1994 Revision*, plus *The Sex and Age Distribution of the World Populations: the 1994 Revision*, *World Urbanization Prospects: the 1994 Revision*, and *Long-range World Population Projections: Two Centuries of Population Growth, 1950-2150*.

- U.S. Bureau of the Census, *World Population Profile: 1994*.

For general surveys:

- World Resources Institute (WRI), *World Resources* (1987 and 1988–89 issues published by Basic Books, New York; 1990–91 through 1994–95 issues cosponsored by the UN Environmental Program (UNEP) and UN Development Program (UNDP) and published by Oxford University Press).

- UN *Statistical Yearbook* (various editions).

- Organization for Economic Cooperation and Development (OECD), *OECD Environmental Data* (Paris: OECD, 1993).

- World Bank, *Social Indicators of Development* (Baltimore: Johns Hopkins University Press, various years).

- UN Food and Agriculture Organization (FAO) "Agrostat" (now "FAO-stat") historical compilations from FAO Yearbooks.

- U.S. Department of Agriculture Economic Research Service (USDA/ERS), *World Agriculture: Trends and Indicators, 1970-91*, Statistical Bulletin 861, November, 1993.

- UN International Labor Organization (ILO), *World Employment 1995* (Geneva: ILO, 1995).

- U.S. Department of Energy Information Administration (DOE/EIA), *International Energy Annual 1993* (U.S. Government Printing Office [GPO], May 1995) and *International Energy Outlook 1995* (June 1995).

- *The World Almanac* (Mahwah, NJ: Funk & Wagnalls, annual).

- Worldwatch Institute *State of the World* and *Vital Signs* annuals (New York: W. W. Norton), which summarize data from the above and other sources and usually get to press faster. Data from all Worldwatch publications are available on annual diskettes (as are most of the series above).

■ ■ ■

For the United States: Bureau of the Census *Population Projections of the United States, by Age, Sex, Race, and Hispanic Origin: 1993 to 2050, Fertility of American Women* (1988 through 1994 editions), *Statistical Abstract of the United States* (various years) and *Historical Statistics of the United States: Colonial Times to the Present, 1957.*

■ Council on Environmental Quality (CEQ), *United States of America National Report* to the United Nations Conference on Environment and Development (UNCED), which is a good summary of environmental issues, as are CEQ's annual and now semi-annual reports *Environmental Quality.*

More specialized reports are put out by the Environmental Protection Agency (EPA) and the Department of Energy. See particularly EPA's *National Air Pollutant Emission Estimates 1940–1990.*

Where data are drawn from other sources and are sufficiently important or contentious to warrant sourcing, they are separately cited.

GEOGRAPHICAL CONVENTIONS

The cold war simplification of "first, second and third world" nations has broken down. The UN in 1991 agreed on two categories of nations: "more developed" and "less developed" countries (LDCs), with the second category containing 47 "least developed" countries, most of them in Africa and Asia. Unfortunately, the world is not that simple, and some "less developed" countries (the "newly industrializing economies" or NIEs) are rapidly passing some "more developed" ones.

Throughout this book, I have used "industrialized" for the nations of the OECD, Russia and the European portion of the erstwhile Soviet Communist bloc. I use the old term "third world" as a nonpejorative shorthand for the rest of the world. I use the term "emerging" or "industrializing" nations to identify the rather amorphous group that is rapidly acquiring some of the characteristics of industrial nations. I call the UN's "least developed" countries "the poorest."

The UN divides Europe into Western, Southern, Northern and Eastern Europe, which includes Russia. These terms are not used in their conventional meanings. The UN definition carries Europe to within two miles of Alaska in the Diomede Islands. My references to "Europe" do not include

Russia. Insofar as possible, I have tried to make statements instead about the Europe that was not Communist, which I label "West Europe" to differentiate it from the UN's smaller "Western Europe."

Many data are now grouped by "EU" (the European Union; the former European Community or "EC" (Ireland, England, Benelux, France, Spain, Portugal, Denmark, Germany, Italy, Greece, and now Sweden, Finland and Austria).

Some data are reported by "OECD" and "non-OECD." The OECD includes the established industrial nations of Europe other than the erstwhile Communist bloc, plus North America (now including Mexico), Japan, Australia and New Zealand.

Energy data still group the "FSU" (former Soviet Union) together. I follow that format in chapter 3.

NOTES

Chapter 1: Population. The Unique Century

1. Leon Bouvier and Lindsey Grant, *How Many Americans?* (San Francisco: Sierra Club Books, 1994).
2. *Overshoot. The Ecological Basis of Revolutionary Change* (Urbana: University of Illinois Press, 1980).
3. Lindsey Grant et al., *Elephants in the Volkswagen* (New York: W. H. Freeman, 1992), p. 142.
4. For a fuller treatment of the ICPD Programme, see Lindsey Grant, "The Cairo Conference: Feminists vs. the Pope," (Teaneck, NJ: Negative Population Growth, Inc., the NPG FORUM series, July 1994).
5. UN Center for Human Settlements, Nairobi, *Global Report on Human Settlements 1995*, reported in "Cities as Disease Vectors," *Science*, November 17, 1995, p. 1125.
6. "NAS Study Highlights Chemical Mutagens," *Science*, March 18, 1993, p. 1304–05, and "Neglected Neurotoxicants," *Science*, May 25, 1990, p. 958.

Chapter 2: The Prospects for Food

7. David Pimentel et al., "Environmental and Economic Costs of Soil Erosion and Conservation Benefits," *Science*, February 24, 1995, pp. 1117–1123.
8. Obaidullah Khan, Assistant Director–General of FAO, Reuters, Bangkok, September 28, 1995.
9. *Science*, July 21, 1995, p. 304, reported high mutation rates among the animals, but growing populations.
10. Norman Myers, "The World's Forests: Need for a Policy Appraisal," *Science*, May 12, 1995, pp. 823–824.
11. "Plants as Chemical Factories," *Science*, May 5, 1995, p. 659.
12. "Toward Sustainable Development," developed by WRI for UNEP and released in March 1992, reported in *Science*, April 3, 1992, p. 28, and summarized in *World Resources 1992-3* (see Note on Sources), pp. 111–118.
13. Gretchen C. Daily, "Restoring Value to the World's Degraded Lands," *Science*, July 21, 1995, pp. 350–358. This estimate, drawn from various studies, is not strictly comparable to the UNEP study above and includes forests and rangeland.
14. Peter M. Vitousek et al., "Human Appropriation of the Products of Photosynthesis," *BioScience*, June 1986, pp. 368–373. This and the Daily study above were part of an ongoing effort to quantify the impact of human activities on resources, led by Paul Ehrlich of Stanford University.
15. This discussion is drawn in part from William C. Paddock, "Our Last Chance to Win the War on Hunger," in *Advances in Plant Pathology* (London, San Diego: Academic Press, 1992), pp. 197–222. Figures for starvation deaths and emigration were taken from the Encyclopedia Britannica 1966.
16. Most of the discussion in this section is drawn from Vaclav Smil, "Population Growth and Nitrogen: An Exploration of a Critical Existential Link," *Population and Development Review*, December 1991, pp. 569–601.
17. Obaidullah Khan. See note 8.
18. Associated Press (AP), Washington, May 21, 1994. AP, Springfield, MA, September 10, 1995.
19. David Pimentel et al., "Environmental and Economic Costs of Pesticide Use," *BioScience*, November 1992, pp. 750–760.

20. Dean Roemmich and Joel McGowan, "Climatic Warming and the Decline of Zoo-plankton in the California Current," *Science*, March 3, 1995, p. 1324, and subsequent "Correction."

21. "Human Appropriation of Renewable Fresh Water," *Science*, February 9, 1996, pp. 785–88.

22. The 10% estimate is taken from Foxworthy and Moody, *National Water Summary 1985* (USGS Water Supply Paper 2300, 1986), pp. 51–68 (courtesy Noel Gollehon, USDA/ERS). The 25% estimate appeared in U.S. Water Resources Council, *The Nation's Water Resources 1975-2000* (U.S. Government Printing Office, 1979), vol. 1 cited in David Pimentel, "Land, Energy and Water" (chapter 2 in *Elephants in the Volkswagen*), and in Gordon Sloggett and Clifford Dickason, *Groundwater Mining in the United States* (USDA/ERS, 1986), cited in Worldwatch Paper, *Full House*, p. 115. Several groundwater specialists at EPA, the USGS and USDA/ERS refused even to make a guess.

23. International Rice Research Institute estimate cited by Reuters, Manila, October 12, 1995.

24. Dr. David Pimentel, Cornell University, personal communication, February 5, 1996.

25. The immediate effect in California may be the wiser use of water since California agriculture has long used heavily subsidized water from the U.S. Bureau of Reclamation, and much water is wasted.

26. AP, Washington, May 6, 1995.

27. Henry W. Kendall and David Pimentel, "Constraints on the Expansion of the Global Food Supply," *Ambio* vol. 23, May 1994, pp. 198–205. See also interview with Dr. Kendall by Nick Ludington, AP, Washington, May 26, 1994.

28. *Report on the Second FAO/UNFPA Expert Consultation on Land Resources for Populations of the Future* (Rome: FAO, 1980).

29. Thomas Malthus, *Population*, the first version, 1798 (Ann Arbor: University of Michigan Press, paperback version, 1959), p. 5. Malthus' essay was directed also at the writings of English philosopher William Godwin.

30. Data assembled from BLM in-house assessments of individual tracts. See Johanna Wald and David Albersworth, *Our Ailing Public Rangelands: Still Ailing! Condition Report—1989* (National Wildlife Federation and Natural Resources Defense Council, October 1989). A summary of nine GAO reports highly critical of BLM management and warning of range deterioration appears in Steve Johnson, "'Disaster, disaster on the range,' Reports Say," in *High Country News*, April 20, 1992, p. 10.

31. A. A. Rosenberg et al., "Achieving Sustainable Use of Renewable Resources," *Science*, November 5, 1993, pp. 828–829.

32. *World Resources, 1994-95*, table 10.1 (see Note on Sources). Population Action International, citing FAO data, puts the current subsidy much higher, at $54 billion. (Washington, D.C.: Population Action International, "Catching the Limits: Population and the Decline of Fisheries," 1995.)

33. The phenomenon is known as "depensation." See "New Study Provides Some Good News for Fisheries," *Science*, August 25, 1995, p. 1043.

34. "Fish, Money, and Science in Puget Sound," *Science*, February 9, 1990, p. 631.

35. See *Science*, February 25, 1994, p. 1089 and August 11, 1995, p. 759. The estimate of plankton's role in absorbing CO_2 is from J. T. Houghton, B. A. Callender, S. K. Varney, eds., *Climate Change 1992: The 1992 Supplementary Report to the IPCC Scientific Assessment* (New York: Cambridge University Press, 1992).

36. "Report Nixes 'Geritol' Fix for Global Warming," *Science*, September 27, 1991, pp. 1490–91.

37. Council on Environmental Quality (CEQ), *Environmental Quality 1993*, p. 34.

Chapter 3: The Energy Transition

38. Charles D. Masters, Emil D. Attanasi and David H. Root, "World Petroleum Assessment and Analysis," *Proceedings of the 14th World Petroleum Congress* (New York: John Wiley & Sons, 1994). These estimates cover recoverable resources of conventional oil and gas, not potential resources such as tar sands. Note that they separately estimate *reserves* already discovered plus *resources* thought to be in place but not proven, with higher and lower bounds for the latter at the 95% confidence level. Data for reserves alone are not of much help, since oil companies usually explore and establish reserves only as needed for business planning purposes.

39. The various projections are spelled out in the Department of Commerce Energy Information Administration (EIA) *International Energy Outlook 1995*. To determine "years of availability," I have run those projections beyond 2010 in the absence of reliable longer term projections. To make the 2.5% growth rate calculation, I applied the EIA data (table 3) as to energy use per dollar of GNP in the non–OECD world to petroleum use. The interpolation is justified since petroleum is expected to retain its approximate proportion of total energy consumption through the EIA projection period.

40. World Resources Institute, UN Environment Programme and UN Development Programme, *World Resources 1994-1995*, op cit., p. 169.

41. Coordination Minister Nygren insisted it must stay (Reuters, Stockholm, September 7, 1995). The ruling Social Democratic Party after a change of premiers voted to stick with plans to phase it out (Reuters, Stockholm, March 17, 1996).

42. Thomas W. Lippman, "Huge Plutonium Stockpile Builds with No Limits in Sight, Study Says," *Washington Post*, November 16, 1993.

43. Italian Nobel laureate Carlo Rubbia, director of the European Laboratory for Particle Physics (CERN), claimed in 1993 (to considerable skepticism) that he had worked out a process to generate energy from thorium, at an efficiency 140 times that of uranium (Reuters, Rome, November 23, 1993).

44. Albert A. Bartlett, "Fusion and the Future," NPG, Inc., *Human Survival*, April 1990, reprinted from the Newsletter of the Forum on Physics and Society.

Chapter 4: The Doomsday Scenarios: Energy, Pollution and Climate

45. CEQ, *Environmental Quality 1993*, pp. 15–16.

46. CEQ, *Environmental Quality 1993*, table 37.

47. *New York Times*, December 23, 1990.

48. See Bouvier and Grant, op cit., p. 85.

49. The National Acid Precipitation Assessment Program, a consortium of representatives from the Departments of Energy, Commerce and Interior, the EPA and the Council on Environmental Quality. The most recent report of its findings is its *1992 Report to Congress*, on which much of the discussion in this section is based.

50. For a summary of the EPA report, see AP, Washington, November 3, 1995. Charles E. Little, *The Dying of the Trees: The Pandemic in America's Forests* (New York: Viking, 1995). See also "Acid Rain's Corrosion Spreads in the Northeast," *Christian Science Monitor*, November 10, 1995, p. 1.

51. "Acid Rain's Dirty Business: Stealing Minerals from the Soil," *Science*, April 11, 1996, p. 198, and G. E. Likens *et al.*, "Long Term Effects of Acid Rain: Response and Recovery of a Forest Ecosystem," ibid., pp. 244–246.

52. News release from the Office of Science and Technology Policy, Executive Office of the President, June 28, 1983.

53. *Science*, April 19, 1991, p. 371.

54. The sources are (a) *International Energy Outlook 1995*, tables A9–A12, (b) *World Resources 1994-5*, Table 23.1, and (c) IPCC, *Climate Change* (see below), pp. 85–86.

55. "IPCC Second Assessment Synthesis of Scientific-Technical Information Relevant to Interpreting Article 2 of the UN Framework Convention on Climate Change, 1995"

(Geneva: World Meteorological Organization, IPCC Secretariat, publication by Cambridge University Press scheduled for 1996). The assessment consists of an overview, plus summary reports from Working Groups I, II and III, supported by literally thousands of pages of documentation. The basis for the conclusions in the summary documents is to be found in other supporting documents, notably R. H. Williams, *Variants of a Low CO2-Emitting Energy Supply System (LESS) for the World* (Richland, WA: Battelle Memorial Institute Pacific Northwest Laboratory, October 1995), plus the 1990 First Assessment report, IPPC, *Climate Change* (Washington: Island Press, 1991).

56. For instance, an old theory has just resurfaced that solar radiance variations of about 0.1% are responsible for global climate changes, though the initial reaction seems to be skepticism that they can cause much of the warming. Dr. Tom M. L. Wrigley of the National Center for Atmospheric Research, a leading participant in the IPCC study, agrees that solar variation may have contributed 10% to 30% to climate warming in the past century. See "A New Dawn for Sun–Climate Links?" *Science,* March 8, 1996, pp. 1360–61.

57. Reuters, London, March 9, 1995.

58. "Monsoon Shrinks with Aerosol Models," *Science,* December 22, 1995, p. 1922. The study was conducted by the Indian Institute of Technology in cooperation with the German Max Planck Institute for Meteorology. A separate study by Britain's Hadley Center came up with comparable results.

59. Robert W. Fox, "Hurricane Andrew: the Population Factor" (Teaneck, NJ: NPG, Inc., NPG Footnote series, September 1992).

60. *Projecting Sea Level Rise,* EPA research paper 230-09-007, October 1983, estimates landbased sea and ice at the equivalent of 78.3 meters' sea level equivalent. Others, adding a factor for thermal expansion of a warmer ocean, use the approximation of 100 meters.

61. Mark Fahnestock, "An Ice Shelf Breakup," *Science,* February 9, 1996, pp. 775–76, and Helmut Rott et al., "Rapid Collapse of Northern Larsen Ice Shelf, Antarctica," same issue, pp. 788–792. (Reuters, Buenos Aires, March 23, 1995; Reuters, London, January 25, 1996.)

62. See, for example, several articles in *Science:* "Greening of the Antarctic Peninsula," October 7, 1994, p. 35; "Is a Warming Climate Wilting the Forests of the North?" March 17, 1995, p. 1595; "Pacific Warming Unsettles Ecosystems," March 31, 1995, pp. 1911–2; "Listen Up! The World's Oceans May Be Starting to Warm," June 9, 1995, pp. 1436–37; Ola M. Johannessen et al., "Global Warming and the Arctic," January 12, 1996, p. 129.

63. Pekka E. Kauppi, "The United Nations Climate Convention: Unattainable or Irrelevant," *Science,* December 1, 1995, p. 1454.

64. "Nyos, the Killer Lake, May be Coming Back," *Science,* June 30, 1989, pp. 1542–42. UPI (Ann Arbor, January 3, 1996.) A group of paleontologists have hypothesized that an enormous "belch" of supersaturated carbon dioxide in the oceans may have contributed to the extinctions at the end of the Permian, 250 million years ago ("Geoscientists Contemplate a Fatal Belch and a Living Ocean," *Science,* December 1, 1995, pp. 1441–42).

65. These ratios reflect the prices paid by Southern California Edison for alternative energy. See Bouvier and Grant, op cit., p. 85. They have the advantage over theoretical projections of being based on real experience; both suggest a serious rise in energy costs with renewables. See for instance Bernard Gilland, "World Population, Economic Growth, and Energy Demand 1990–2100" in *Population and Development Review,* September 1995, pp. 507–541.

66. Joel E. Cohen catalogues the different estimates through the years in *How Many People Can the Earth Support?* (New York: W. W. Norton, 1995) but does not himself offer an explicit answer.

67. James Kasting (Pennsylvania State University) and James Walker (University of Michigan), cited in "No Way to Cool the Ultimate Greenhouse," *Science*, October 29, 1993, p. 648.

Chapter 5: Technology: Deus ex Machina?

68. Reuters, Geneva and Washington, February 14, 1995.

69. *Science*, July 21 and September 1, 1995. There is a debate raging as to whether UV radiation is the sole cause of the decline of amphibians but no doubt expressed that it is a cause.

70. Reuters, Harare, September 25, 1995. Perhaps I should not be so hard on the Africans. Florida fruit and vegetable growers won a reprieve from 2000 to 2001 in phasing out the fumigant—as one of the deals to gain U.S. Congressional approval for NAFTA, according to critics (AP, Washington, November 30, 1993).

71. John L. Peterson (The Arlington Institute), *The Road to 2015* (Corte Madera, CA: The Waite Group Press, 1994), p. 76, citing J. O. Nriagu, "Global Metal Pollution," *Environment* 32 (7), pp. 7–32.

72. *Technical Bases for Yucca Mountain Standards*, report of a National Research Council study on health standards for nuclear waste repositories (Washington: National Academy Press, 1995) reported in "Taking the Long View of Yucca Mountain," *Science*, August 11, 1995, p. 759.

73. The present model runs on methane, which still generates CO_2, though less than conventional engines. A pure hydrogen engine is still a decade or more away (Reuters, Ulm, Germany, April 13, 1994).

74. G. Pascal Zachary, "Worried Workers. Service Productivity Is Rising Fast—and So Is the Fear of Lost Jobs," *Wall Street Journal*, June 8, 1995, pp. A1, A10.

75. Leontief even disconnected productivity from the labor supply in a multisectoral model of the world economy done with Anne Carter of Brandeis University in 1977 for the UN. See *The Future of the World Economy* (New York: Oxford University Press, 1977) and Leontief and Ira Sohn, "Population, Food and Energy and the Prospects for Worldwide Economic Growth to the Year 2030" (Paper for the Nobel Symposium on Population Growth and World Economic Development, Oslo, Norway, September 7–11, 1981). Jeremy Rifkin (President, Foundation on Economic Trends, Washington), *The End of Work: The Decline of the Global Labor Force and the Dawn of the Post-Market Era* (New York: Tarcher–Putnam, 1995) makes the same points with considerably more documentation and calls for government support of "third sector" jobs in lieu of welfare, along lines similar to my proposal in chapter 12.

76. Rich Miller, "World Economy Faces Jobs Crisis, UN Agency Says," Reuters, Washington, February 2, 1994, phrases it as "not earning a minimal subsistence," quoting a new ILO report titled "Defending Values, Promoting Change."

77. Chief UN arms inspector in Iraq, per Reuters, UN, August 16, 1995. The reference to radiological weapons is from a senior scientist at Lawrence Livermore Laboratory, per UP, Livermore, CA, November 10, 1995.

78. "The Defense Initiative of the 1990s," *Science*, February 24, 1995, pp. 1096–1100.

79. AP, Columbus, June 6, 1995. AP, Little Rock, December 23, 1995.

80. "Discovering Microbes with a Taste for PCBs," *Science*, October 28, 1987, pp. 975–77, and "Bacteria Effective in Alaska Cleanup," *Science*, March 30, 1990, p. 1537.

81. See, for instance, the Department of Energy sponsored work by United Solar Systems Corp. in developing a low–cost amorphous silicon film for photovoltaic energy production, said to be 10% efficient and to cut the costs of photovoltaics in half. It is claimed that production will begin momentarily ("New Material Could Halve Solar Costs," *Washington Post*, January 18, 1994).

Chapter 6: A Conserving State of Mind

82. Lynn Margulis, Professor of Biology at Boston University, "Gaia, A New Look at the Earth's System," in *Technology, Development and the Global Environment* (Mahwah, NJ:

Ramapo College Institute for Environmental Studies, 1991), pp. 299–305. See also Charles Mann, "Lynn Margulis: Science's Unruly Earth Mother," *Science*, April 19, 1991, p. 378, and her response "Gaia" in *Science*, February 5, 1993, p. 745.

83. "Uganda Enlists Locals in the Battle to Save Gorillas," *Science*, March 24, 1995, pp. 1761–62. "Gorillas Killed in Ugandan Park," *Science*, April 7, 1995, p. 25.

84. Norman Myers, "Environmental Unknowns," *Science*, July 21, 1995, pp. 358–360.

Chapter 7: Diverging Futures

85. ILO annual reports 1994 and 1995.

86. 1994–95 WRI annual (see Note on Sources), figure 6.2. Note that this figure includes North Africa; the decline in sub–Saharan Africa is probably even steeper.

87. AP, Washington, October 20, 1995.

88. E. O. Wilson, *The Diversity of Life* (New York: Norton, 1992), pp. 328–29.

89. United Nations, *World Population Prospects: the 1992 Revision*, chapter 3. The countries are Malawi, Rwanda, Uganda and Zambia. One cannot draw even from this catastrophe the bitter hope that AIDS may slow Africa's population growth rate. Not yet. The UN projects that as a result of the epidemic, the annual population growth of the 15 most affected nations (all in Africa), from 1995–2000 will be 2.96% rather than 3.26%. At 2.96%, it still takes just 24 years for populations to double. Expect more trouble.

90. Worldwatch, *Vital Signs, 1994*, cites the Global AIDS Policy Coalition at the Harvard School of Public Health for the 5 million AIDS figure, but a UN report prepared by the Bagnoud Center for Health and Human Rights at the same school recently used an estimate of 1.4 million. See AP, Geneva, July 6, 1996, and Reuters, Vancouver, July 8, 1996. The U.S. Census Bureau uses an intermediate estimate of 2.5 million.

91. John Larner, *Culture and Society in Italy 1290-1420* (London: B.T. Batsford, 1971). See pp. 122–145 and 237–240 for the direct evidence; the discussion of Changes in Art, pp. 141–147 is suggestive of the intellectual impact. See also John A. Garraty and Peter Gay, *The Columbia History of the World* (New York: Harper & Row, 1972), p. 487.

92. Peng Peiyun, Minister of the State Family Planning Commission, April 21, 1993, quoted in *Population Today*, June 1993, p. 8.

93. The worst excess apparently is that of overzealous local officials forcing late–term abortions on women in order to stay within quotas. The recent report by Human Rights Watch (AP, Beijing, January 6, 1996) of orphanages starving the children probably reflects official loss of control over expanding corruption. Pilfering from food budgets has a long history in penal institutions and orphanages, worldwide. China probably has not deliberately extended its demographic policies to raise mortality. There is a suspiciously high proportion of males in the official birth statistics, suggesting some female infanticide and/or parents' failure to report the births of girl children in order to try again for a boy. However, child mortality is extremely low in China by comparison with other countries with similar per capita income.

94. Official Chinese statistics are frequently in conflict, underlying the danger of too close a reliance upon detailed data (see Introduction). The official Xinhua News Agency quotes a spokesperson for the State Land Administration: "China has 120 million hectares of cropland, not just 100 million hectares as long reported." The Minister of Agriculture refused to comment (Reuters, Beijing, March 10, 1995). The larger figure is credible, since peasants and local and provincial governments all have reason to understate acreage to avoid forced deliveries of grain to the center. A 20% discrepancy suggests that inferred fertilizer use and crop yields per hectare are substantially less than the official figures. Perhaps China has a bit more room for improvement than official data suggest.

95. USDA data and Reuters, Beijing, August 29, 1994, quoting the Director General of the State Land Administration.

96. Reuters, Beijing, July 19, 1995.
97. AP, Beijing, September 11, 1995, quoting the official Xinhua News Agency.
98. The period cited was April 1994 to April 1995. Reuters, Peking, May 15, 1995. The Worldwatch Institute uses a figure of 60% for 1994.
99. Data, including 1994 price change, and projections are from Lester R. Brown, *Who Will Feed China?* (New York: W. W. Norton, 1995) pp. 96–97.
100. Reuters, Beijing, March 5, 1995, and August 3, 1995, quoting Xinhua News Agency and experts from the Chinese Academy of Social Sciences and Academy of Agricultural Sciences, who estimated the shortfall at 50 million tons. A group of Chinese and Japanese economists arrived at a shortfall of 136 million tons in 2010 (Reuters, Bangkok, September 29, 1995).
101. Reuters, Beijing, October 16, 1995, quoting the Deputy Director for Commerce and Industry, State Planning Commission.
102. The energy program director at the East West Institute, Honolulu, quoted by Reuters, Singapore, September 16, 1994.
103. Reuters, Beijing, October 4, 1995, report on the draft Plan. The employment figures below are drawn from this release and from a Reuters, Beijing, May 29, 1995, summary of a report by a Central Committee researcher quoted in the *Economic Daily*.
104. Reuters, Beijing, January 5, 1994, quoting the Minister of Agriculture and a provincial labor official, respectively.
105. The proposal came from Shenyang, in the northeast, and was cited approvingly by the official New China News Agency (Reuters, Beijing, July 31, 1993).
106. The World Bank, U.S. Bureau of the Census and United Nations projections involve different assumptions as to what China's fertility has been and will be, leading to different conclusions as to the pace of decline.
107. Premier Li Peng to the World Bank President (Reuters, Beijing, September 20, 1995).
108. *Beijing Review*, November 6–12, 1995.
109. AP, Beijing, via Compuserve Executive News Service, September 30, 1995. The figure cited was 6.5 million cubic yards. The figure presumably includes industrial sewage.
110. Masayoshi Sadakata of Tokyo University, quoted in *Science.* July 21, 1995, p. 296.
111. Open Media Research Institute, Prague, "Cabinet Ministers Express Concern Over Illegal Chinese Immigration," August 25, 1995.
112. UN World Commission on Environment and Development (or "Brundtland Commission" for its Chair, Gro Harlem Brundtland, Prime Minister of Norway), *Our Common Future* (Oxford; Oxford University Press, 1987), p. 213.
113. U.S. Department of Energy (DOE), Energy Information Administration (EIA), *International Energy Annual 1993*, p. viii.
114. EIA, *International Energy Annual 1993*, table WEO2.
115. EIA, *International Energy Outlook 1995*, figure 3.
116. Reuters, Seoul, November 15, 1994.
117. Grant et al., op cit., chap. 17.
118. Reuters, London, July 28, 1995.
119. Smil, op cit. (note 16), p. 589.
120. In the UN classification, Northern Europe includes the United Kingdom, Ireland, Scandinavia and the Baltic states; Western Europe includes Austria, Germany, France, the Benelux countries and Switzerland; Southern Europe includes the remainder of Europe other than Eastern Europe (the former Soviet bloc European countries other than the Baltic states).
121. Marion Donhoff et al., "Damit die Deutschen nicht Aussterben," in *Weil Das Land Sich Andern Muss* (Reinbeck bei Hamburg: Rowohlt Verlag, 1992). Translated into English as "Saving the Germans from Extinction," Jack Miles, translator, *The Social Contract*, Fall 1995.
122. AP, Washington, September 24, 1993, 17:57 EDT.

123. Reuters, Washington, September 24, 1993.
124. Alan Wheatley, "OECD Fears Social Explosion from Unemployment," Reuters, Paris, June 1, 1993.
125. Reuters, Paris, June 18, 1993.
126. Alan Wheatley, "Unions Say 'Yes, but' to OECD's Jobless Cure," Reuters, Paris, June 3, 1993.
127. Reuters, Paris, January 23, 1996.

Chapter 8: One World, Like It or Not

128. See Bouvier and Grant, op cit., pp. 43–45, for a fuller discussion of the resurgence of epidemics in recent years.
129. Robert W. Fox, "Neighbors' Problems, Our Problems," Chapter 18 in Lindsey Grant et al., *Elephants in the Volkswagen* (New York: W. H. Freeman, 1992). The figures do not include Mexico.
130. Reuters, Rome, January 29, 1996, citing the Chairman of the Food Security Committee of FAO.
131. Reuters, Dhaka, September 20, 1995.
132. Voice of America (VOA), Brussels, September 18, 1995, citing a European Union Secretariat study.
133. Reuters, London, September 21, 1995.
134. AP, Washington, February 27, 1996.
135. This total includes Japan, 1989–91 average; the figure differs from the conventional 200 million ton figure because it nets out some intraregional trade.
136. Reuters, Rome, September 20, 1995.
137. UN Centre on Transnational Corporations, *Transnational Corporations in World Development, Third Survey* (New York: UN, 1983).
138. AP, Geneva, April 10, 1995.
139. Clay Chandler, "Rubin Faces Wary Audience on Free Trade; Pacific Rim Nations' Visions Vary." *Washington Post*, April 16, 1995.
140. John B. Cobb, Jr., and Herman Daly, "Free Trade vs. Community: Social and Environmental Consequences of Free Trade in a World with Capital Mobility and Overpopulated Regions," in *Population and Environment*, Spring 1990.
141. Joseph Kahn, "China Swiftly Becomes an Export Colossus, Straining Western Ties," citing Chinese data. *Wall Street Journal*, November 13, 1995, p. 1.
142. U.S. International Trade Commission, "The Likely Impact on the United States of a Free Trade Agreement with Mexico" (Washington, February 1991), pp. 2–5.
143. Robert Repetto and William B. McGrath, *Wasting Assets: Natural Resources in the National Income Accounts* (Washington: World Resources Institute, 1989). This was a pioneering effort. WRI subsequently published a Repetto study on Costa Rica titled *Accounts Overdue: Natural Resource Depletion in Costa Rica* (1992). Several countries, plus the UN, have studies under way to develop national income accounting systems that take account of resource depletion, and some even include efforts to factor quality of life into such accounts. See "Accounting for the Environment," *Science*, December 20, 1991, p. 1724, and the proposal for a Index of Sustainable Economic Welfare in Herman Daly and John B. Cobb, Jr., *For the Common Good* (Boston: Beacon Press, 1989).
144. Reuters, Berlin, March 26, 1995.
145. This account is taken from wire service reports, notably Reuters, Washington, March 17, 1995; Reuters, Dubai, April 2, 1995; Reuters, Berlin, March 26, 1995, March 27, 1995, and April 7, 1995; Reuters, New Delhi, April 5, 1995; and AP, Berlin, April 10 and 12, 1995.
146. The 1995 statement was an outgrowth of one authored in 1966 by John D. Rockefeller III. The present version was circulated by Population Communication (Los Angeles), with the help of the Population Institute (Washington, D.C.).

147. The Department's Israel Desk states that the United States is financing water projects in the Gaza Strip and plans to do so on the West Bank. At a forthcoming multilateral Palestinian assistance conference in Paris, it plans to announce over $50 million of such projects. There is a "water working group" among the five multilateral groups assisting the Palestinians. Neither the Israel Desk nor the Near Eastern Bureau could offer any instances in which the United States has suggested that there is a connection between water scarcity and population growth (telephone conversations December 13, 1995). Secretary Warren Christopher subsequently made that connection, but went on only to say that the United States is helping the nations of the region to manage their water resources, not to manage population growth (Stanford University speech, April 9, 1996). To his credit, later in the same speech he said, "we must ease the pressures of deforestation and rapid population growth" in Haiti. He alluded to cooperation with Japan to "stabilize population growth" internationally. (It's the population and not the growth that must be stabilized—or better—but he is on the right track.)

148. The director of the Palestine Bureau of Statistics, quoted by AP, Al Beireh, West Bank, February 6, 1996.

149. The inequitable water distribution is confirmed by a University of Toronto/American Academy of Arts and Sciences study analyzing the connections between resource depletion and violent conflict. See Thomas F. Homer-Dixon, Jeffrey H. Boutwell and George W. Rathgens, "Environmental Change and Violent Conflict," *Scientific American*, February 1993, pp. 38–45. The quotation about the Gaza strip is from Reuters, Amman, May 31, 1994, reporting on a meeting of the UN Economic and Social Commission for Western Asia (ESCWA).

150. See Reuters, Muscat, dispatches April 15, 18, and 19, 1994, reporting the first regional water conference, hosted by Oman. See also Sandra Postel, *Last Oasis: Facing Water Scarcity* (Washington: Worldwatch Institute, 1992).

151. AP, Jerusalem, October 9, 1993.

152. In addition, of course, there is the cost of raising and moving the water from the sea to the place it is needed. A recent report (Reuters, Nicosia, January 31, 1996) described a major desalinization plant to be built for the Government of Cyprus, which will pay a charge of only $1 per cubic meter for the water delivered, but this appears to be additional to the capital costs.

Chapter 9: U.S. Population Growth: An Accidental Future

153. Leon F. Bouvier and Lindsey Grant, *How Many Americans?* (San Francisco: Sierra Club Books, 1994). This projection is conservative both as to immigration and fertility, using an old Census Bureau figure for net immigration that is almost certainly below present levels, and assuming that immigrants' fertility drops substantially in the second generation. The Census Bureau projection below is the new 1996 variant, and its middle projection drops even that level to 820,000 per year.

154. In this projection, I have assumed constant fertility and mortality and used the Census Bureau's "high" figure of 1,370,000 of anticipated annual net immigration, which I believe is probably closer to present reality than its "medium" figure of 880,000.

155. Lindsey Grant, "What We Can Learn from the Missing Airline Passengers" (Teaneck, NJ: NPG, Inc., NPG FORUM series, November 1992). Overall, about 2 million people more arrive by air than are reported departing. Less than half the gap can be explained by official immigration figures. I suggested that if the data are even approximately right, there are far more people overstaying their visas than the official statistics suggest. If the data are wrong, it would be relatively easy to collect better data by enforcing the rules requiring departing aircraft to file manifests. The Census Bureau uses these residual figures to estimate movement to and from Puerto Rico but not more generally, and nobody seems anxious to improve the data. It is rather foolish to spend money to collect data that you don't bother to use.

156. Center for Immigration Studies, Washington, D.C., *Immigration Statistics - 1994* and *1995*. CIS Backgrounders 1–94 (May 1994) and 2–95 (July 1995).

Chapter 10: Living with the Land

157. AP, Washington, January 25, 1996.
158. See "Can Sustainable Farming Win the Battle of the Bottom Line?" *Science*, June 15, 1993, pp. 1893–1895, for a useful summary of the U.S. experience.
159. From a State of New Mexico State Engineer Office unpublished memorandum titled "Water Conservation Program Update," dated February 10, 1995.
160. The revised estimates (which do not include federal offshore fields) raise the 1989 estimated reserves/resources by 43% and the 1994 figures by 20%. This gain reflects new data on "inferred reserves" in existing fields, the inclusion of "unconventional resources," new extraction technologies, and a relaxed definition of reserves, from "economically recoverable" to potentially recoverable in the next half–century with "a reasonable price increase." The new inferred reserves are described as far from exploitable at today's prices, but the authors expect prices to rise. The estimate of undiscovered resources fell dramatically in the 1980s. See USGS Survey Circular 1118, *1995 National Assessment of United States Oil and Gas Resources*, as amplified by the misleadingly labeled "U.S. Oil and Gas Fields Double in Size," *Science*, February 24, 1995, pp. 1090–92, and by telephone conversation with Ronald Charpentier, USGS Denver, March 7, 1995.
161. The comparisons in this section are taken from the *U.S. Statistical Abstract, 1994*, tables 920, 1366, 1370 and 1393 and earlier issues. GNP is in constant dollars per capita, energy use per capita, and the comparison is adjusted (table 1370) to reflect the somewhat lower real GNP per capita in Japan, in terms of purchasing power.

Chapter 11: The Collapsing Society

162. Roy Beck, *The Case Against Immigration* (New York: W. W. Norton, 1996), chapter 10.
163. *Statistical Abstract of the United States*, tables 308–324. I use the number of victims as a measure, deliberately, because it is likely to be the most accurate count and is not subject to charges of bias.
164. AP, Washington, June 2, 1994.
165. Steven Rattner (a managing director of Lazard Freres & Co.), "GOP Ignores Income Inequality," *Wall Street Journal*, May 23, 1995. He called income inequality "the greatest evident threat to America's well-being."
166. AP, Washington, August 14, 1995, citing the Luxembourg Income Study led by two U.S. professors and financed by the National Science Foundation, among other contributors.
167. Edward N. Wolff, "Top Heavy" (Twentieth Century Fund, quoted by Arthur Schlesinger, Jr., "In Defense of Government," *The Wall Street Journal*, June 7, 1995).
168. Dale Russakoff, "No-Name Movement Fed by Fax Expands; Political Networks Combine Technology, Fear" and Tamar Jacoby, "Conservative, African American, Making Waves," both in *Washington Post* via Compuserve, August 19, 1995.
169. The consulting firm Challenger, Gray & Christmas runs periodic surveys and counted 515,292 jobs lost in 1991, 615,186 in 1993 and 516,069 in 1994. They estimate that they missed 20% more. Quoted by Frank Swoboda, "U.S. Companies Speed Pace of Downsizing" (*Washington Post*, February 8, 1994) and Matt Murray, "Thanks, Goodbye. Amid Record Profits, Companies Continue to Lay Off Employees" (*Wall Street Journal*, May 4, 1995, p. 1).
170. AP, New York, September 17, 1995.
171. The Department of Energy has announced a 27% cut in its work force (*Wall Street Journal*, May 4, 1995), NASA and DOD are engaged in massive reductions.
172. Keith Bradsher, "Skilled Workers Watch Their Jobs Migrate Overseas: A Blow to Middle Class," *New York Times*, August 28, 1995, p.1.

173. Timothy J. Hatton and Jeffrey G. Williamson, "What Drove the Mass Migrations from Europe in the Late Nineteenth Century?" *Population and Development Review*, September 1994, pp. 533–560.

Chapter 12: Population and Policy

174. For the derivation of the "Foreign Entrants" column, see my NPG FORUM article, "A Beleaguered President, A Fizzled 'Economic Stimulus Package,' and a NAFTA Time Bomb," May 1993, note 12. The Bureau of Labor Statistics estimates that 57% of new immigrants, refugees and legalized alien residents are in the work force plus all of the nonimmigrants who apply for temporary work permits. The estimate is necessarily imprecise. Some of those legalized or given work permits were probably already holding jobs. On the other hand, the figure does not include FY1992 illegal border crossers and those on nonimmigrant visas working illegally. Job creation data from *U.S. Statistical Abstract, 1993*. The INS in 1995 announced a policy requiring a waiting period before asylum applicants can get work permits, but as the graph makes clear, these are only a small part of the problem.

175. John E. Rielly, ed., *American Public Opinion and U.S. Foreign Policy* (The Chicago Council on Foreign Relations, 1991 and 1995). There are minor internal discrepancies presumably the result of rounding. I have used figure VI-1 in those cases.

176. Herbert Rowen, "How to Create a Million New Jobs," *Washington Post*, November 15, 1993. Rohatyn advocated a huge public works program to be financed with a gasoline tax.

177. In August 1996 the President signed a complex welfare reform bill transferring management of several federal welfare programs to the states, terminating those programs for any given recipient after two years, and putting a five-year cap on welfare for most individuals. Five months earlier, Congress with Presidential connivance had just shot down the first serious effort in years to bring immigration under control (chapter 17). Washington has tried to cure the symptoms without treating the disease. What if there are no jobs? The reform is illusory, but it will serve both parties in the 1996 election campaign.

178. AP, Washington, February 5, 1994. Reuters, Washington, September 8, 1993.

179. AP, Washington, June 1, 1993.

180. "Fixing the Welfare Mess," *U.S. News & World Report*, December 13, 1993, quoting "a recent study by LaDonna Pavetti."

181. AP, Washington, January 30, 1994.

182. AP, Washington, January 30, 1994.

183. These ideas are sketched out more fully in "Into the Wind . . . Unemployment and Welfare Reform" (NPG FORUM series, March 1994).

184. George Borjas, "The Welfare Magnet," *National Review*, March 11, 1996, pp. 48–50.

185. See Lindsey Grant, "Demography and Health Care Reform" (Teaneck, NJ: NPG Inc., NPG FORUM, October 1993) for fuller treatment.

186. Joseph Califano, "The Last Time We Reinvented Health Care," *Washington Post*, OpEd page A23, April 1, 1993.

187. See Lindsey Grant et al., *Elephants in the Volkswagen* (New York: W. H. Freeman & Co., 1992), Chapter 12, "Too Many Old People?"

188. Annual figures from *Statistical Abstract of the U.S., 1992*, table 192. An unknown part of the growth resulted from improved reporting. Other data through March 31, 1993, by telephone on May 28, 1993, from HHS Centers for Disease Control, Atlanta.

189. Karen Hein, director of the adolescent AIDS program, Montefiore Medical Center, to the Ninth International Conference on AIDS, quoted by Cynthia Johnson, "Teenagers Seen Leading Next Wave of AIDS Epidemic," Reuters, Berlin, June 5, 1993.

190. AP, Geneva, June 3, 1995, quoting the World Health Organization.

191. Dr. Robert Pinner of the Centers for Disease Prevention and Control, cited by Reuters, Washington, January 16, 1996.

192. For a catalogue of diseases, see Arno Karlen, *Man and Microbes: Disease and Plagues in History and Modern Times* (New York: Putnam, 1995). See also Daniel E. Koshland, Jr., "The Microbial Wars," editorial in *Science*, August 21, 1992, p.1021. National Institute of Medicine, *Emerging Infections: Microbial Threats to Health in the U.S.* (Washington: National Academy Press, 1992). American Society of Microbiology statement from AP, Washington, June 5, 1995.

193. AP, Washington, August 23, 1993.

194. Reuters, Los Angeles, August 13, 1993.

195. Surgeon General Joycelyn Elders, herself Black, said that Medicaid must have been "developed by a white male slave owner," because nobody else would want to encourage the proliferation of healthy, pregnant and unemployed young women. AP, Washington, February 25, 1994.

196. Reuters, Nicosia, June 3, 1993.

Chapter 13: The Population Solution

197. The TFR would actually be 1.495. One can make the calculation from table 1 of *Fertility of American Women* (Note on Sources). The oldest cohort (40–45 years old) has a (substantially) completed TFR of 1.999, of which 0.504 represents third order children and above. The residual is 1.495.

198. The "unplanned" projection is from Bouvier and Grant, op cit., Chapter 2. The "planned" projection was run by Decision Demographics, a branch of the Population Reference Bureau, at the request of Negative Population Growth, Inc., in 1992, based on the 1990 population level and assuming a TFR of 1.5 starting from that date. It is the basis for the calculations in NPG Position Paper "Why We Need a Smaller Population and How We Can Achieve It," July 1992.

199. Bouvier and Grant, op cit., table 4.4.

200. *How Many Americans?* op cit., table 4.3.

201. U.S. Bureau of the Census, *Fertility of American Women: June 1992*, Publication P20–470, June 1993, tables 1 and 10. The proportion of one- and two-child families among Blacks and Hispanics is somewhat lower: 62% and 50%, respectively. This probably is a reflection of economic status and education more than of race, but data are not available to check that assumption. Expected number of children is almost identical for Blacks and Whites, but still somewhat higher among Hispanics.

202. See John R. Weeks, "How to Influence Fertility: the Experience So Far," in the NPG FORUM series, 1990, or in Lindsey Grant et al., *Elephants in the Volkswagen* (New York, W. H. Freeman & Co., 1992) Chapter 15.

203. (a) The 1988–94 fertility surveys are averaged to provide a meaningful sample size. (b) "SomeColl" includes the final column "GradDegree." These groups are self-identifying, and the "GradDegree" column in the 1988 and 1990 surveys include all those with five or more years of college. (c) Hispanics are asked to identify themselves as either Black or White, and thus are included in the first two columns and separately reported in the third. This overlap is corrected in some Census statistics but not the fertility reports. This may change; Census is contemplating dropping the dual identification in its 2000 Census. (d) The apparent high fertility of Hispanic women with graduate degrees may be a product of the sample size. Although more than 200,000 women were interviewed for the four surveys, the total number of Hispanic graduate degree holders was only 448, and their fertility varied widely from survey to survey.

204. The application of natural selection to human populations is described at length in biologist Garrett Hardin's *Living Within Limits* (Oxford and New York: Oxford University Press, 1993), Chapter 16.

205. Greg Burke, "Pope Warns UN Not to Try to Limit Family Size," Reuters, Vatican City, March 18, 1994. For more on the Vatican position preparatory to the ICPD, see Philip

Pullella, "Vatican Summons Envoys Over Population Issues," Reuters, Vatican City, March 25, 1994.

206. Reuters, Los Angeles, August 9, 1993, 18:55 and United Press (UP), September 3, 1993.

207. For a more detailed discussion, see Lindsey Grant, "The Cowering Giant" (NPG, Inc., the NPG Footnote series, October 1995).

208. See for example David Simcox, NPG FORUM paper "Sustainable Immigration: Learning to Say No" (1990) and Simcox and Rosemary Jenks, NPG FORUM paper "Asylum Policy: National Passion vs. National Interest" (1992), plus the Federation for American Immigration Reform (FAIR), *Ten Steps to Ending Illegal Immigration* (Washington: FAIR, 1995).

209. AP, Washington, February 5, 1994.

210. AP, Washington, December 29, 1994.

211. School of Public Affairs, University of Maryland, "Americans and Foreign Aid: A Study of American Public Attitudes," January 1995, reported in Population Reference Bureau, *Population Today*, April 1996.

212. For the calculations, see the UN ICPD Secretariat "Background Note on the Resource Requirements for Population Programmes in the Years 2000–2015," June 23, 1994. The direct family planning needs are substantially smaller, since these figures are inflated with expenditures on "reproductive health" and AIDS prevention. For fuller discussion, see my NPG FORUM paper "The Cairo Conference: Feminists vs. the Pope" (Teaneck, NJ: NPG, Inc., July 1994).

Chapter 14: The War of the Paradigms

213. Orville L. Freeman, "Meeting the Food Needs of the Coming Decade," *The Futurist*, November/December 1990, p. 16.

214. Earth Day speech 1993.

215. AP, Kaunas, Lithuania, September 6, 1993.

216. Herman E. Daly and John Cobb, Jr., *For the Common Good* (Boston: Beacon, 1989), chapter 20.

217. Alden Speare, Jr., and Michael J. White, "Optimal City Size and Population Density for the Twenty-First Century," in *Elephants in the Volkswagen*, pp. 94–95.

218. Grant et al., op cit., p. 10.

Chapter 15: Of Tigers, Ants and People

219. Clarence Day, Jr., *This Simian World* (New York: Knopf, 1920).

220. Keynes is quoted by E. F. Schumacher, *Small is Beautiful. Economics as if People Mattered* (New York: Harper & Row, 1973), p. 24.

Chapter 16: Multiple Agendas and the Population Taboo

221. National Audubon Society, *Population Newsletter*, Summer 1994, p. 5.

222. UNFPA, *Populi*, May 1994, p. 10.

223. ICPD Newsletter No. 14, April 1994, p. 7.

224. For a detailed discussion of the Cairo Conference, see NPG FORUM paper "The Cairo Conference: Feminists vs. the Pope," July 1994. That paper analyzed the draft Programme, but the final version is substantially unchanged except for the use of compromise language on abortion.

225. *Sierra*, September/October 1994, p. 52.

226. Mark Mardon in *Sierra*, January/February 1994, pp. 12, 14. He was responding to criticisms of his review of a book by Garrett Harden.

227. For a refutation of the common argument that immigrants do not supplant U.S. labor, see Donald L. Huddle, "Immigration, Jobs & Wages: The Misuses of Econometrics" (NPG FORUM, April 1992). For a report of an effort to measure the displacement in one city, see his NPG FORUM paper, "Immigration and Jobs: the Process of Displacement" (NPG FORUM, May 1992).

228. National Audubon Society and Population Crisis Committee, *Why Population Matters. A Handbook for the Environmental Activist*, 1991, p. 5.

229. See Lindsey Grant, "Free Trade and Cheap Labor: the President's Dilemma" (NPG FORUM paper, October 1991). The $200 billion assumes a cost of $80,000 per job, with jobs for everybody age 15–64. In the capital–intensive United States, business' total investment works out to $115,000 per job, but this does not include the construction and maintenance of social infrastructure, which is difficult to correlate directly with job creation but since 1960 has absorbed close to 3% of U.S. GNP annually.

230. David Simcox, *The Pope's Visit: Is Mass Immigration a Moral Imperative?* (NPG Inc., NPG Footnotes series, December 1995).

231. AP, Montreal, October 31, 1995.

Chapter 17: The Failure of Leadership

232. Kennedy P. Maize, ed., *Blueprint for the Environment* (Salt Lake City: Howe Brothers, 1989), pp. 6, 28–29.

233. The following discussion is a condensation of my NPG FORUM paper "The Timid Crusade," which documents the failure of major environmental groups to support meaningful proposals to reduce population growth. In "major environmental groups," I included Defenders of Wildlife, Friends of the Earth, Greenpeace, the National Audubon Society, the National Parks and Conservation Association (NPCA), the National Wildlife Federation, the Natural Resources Defense Council (NRDC), the Population Institute, the Sierra Club, the U.S. chapter of the World Wildlife Fund, and The Wilderness Society, plus a major population group, Zero Population Growth (ZPG). On February 23, 1996, The Wilderness Society adopted a new policy statement that concluded that "both [U.S.] birth rates and immigration rates need to be reduced." There are signs of change in ZPG. The Sierra Club, on the other hand, has moved backwards and adopted a policy that it will not address immigration issues.

234. *Earth Island Journal*, Spring (Northern Hemisphere), Fall (Southern Hemisphere) (sic), 1994.

235. For example, the David and Lucile Packard Foundation "has made a decision long ago to not work in the area of immigration. A project that includes the impact of developing countries' growth on the United States is therefore, unfortunately, outside our grant guidelines." Letter dated May 16, 1995, from Martha Campbell, Program Officer, Population, Packard Foundation. The Foundation is trying to work with those countries on family planning programs.

236. U.S. President Richard M. Nixon, "Special Message to the U.S. Congress on Problems of Population Growth, July 18, 1969," in *Public Papers of Presidents of the United States* (Washington: Office of the Federal Register, 1969, p. 521).

237. *Population and the American Future. The Report of the Commission on Population Growth and the American Future* Summary volume. (New York: Signet Books, 1972), pp. 76, 184–190 and 204–206. For a summary of the recommendations and a follow–up report on what has happened, see David Simcox, "The Commission on Population Growth and the American Future. 20 Years Later: A Lost Opportunity" (*The Social Contract*, Summer 1992, pp. 197–202).

238. U.S. Council on Environmental Quality and Department of State, *Global Future: Time to Act* (U.S. Government Printing Office, January 1981), p. 11.

239. The President's Council on Sustainable Development (PCSD), *Sustainable America. A New Consensus* (U.S. Government Printing Office, February 1996). An evaluation of the PCSD in midcareer is contained in NPG FORUM paper, "Sobering News from the Real World" (February 1995) and a final brief summary of the population aspects in NPG Booknote, "Population and the PCSD" (April 1996).

240. Susan Chira, "Women Campaign for New Plan to Curb the World's Population," *New York Times*, April 13, 1994, pp. A1, A12.

241. A sampling of Roper Polls in 1977, 1980, 1985 and 1995 showed majorities from 62% to 80% favoring reduced legal immigration and 89% to 91% calling for more vigorous enforcement of our immigration laws (private communication). A survey of polls by the Communications Consortium Media Center, "A Summary of Public Opinion on Immigration" (done in 1994 for the Pew Global Stewardship Initiative, Washington D.C.), showed similar results. The poll of Hispanics was the Latino National Political Survey report "Latinos Speaking in Their Own Voices," released December 15, 1992, table 7.24 (Multi-university research project jointly funded by the Ford Foundation and a consortium of the Rockefeller, Spencer and Tinker foundations). An in-depth survey co-sponsored by the Ford Foundation (which itself funds pro-immigration groups) reached the same conclusion, showing majorities of 66% (Cuban) to 79% (Puerto Rican) among different Hispanic groups favoring reduced immigration.

242. "Population Attitude Study December 1995 for Negative Population Growth, Inc." Released February 17, 1996. The poll also showed considerably more support for stopping population growth than had been anticipated: 59% wanted it stabilized at or below today's level, and 81% wanted it stabilized below 400 million, which is where we are heading by 2050 with present laws.

243. AP, Washington, April 26, 1996.

244. The amendment is dissected in "Protecting American Workers?" (Center for Immigration Studies *Announcement*, April 10, 1996).

245. William Claiborne, "Administration Unveils Plans to Beef Up Border Control," *Washington Post*, February 3, 1994.

246. AP, Washington, September 13, 1995, and conversation with the Center for Immigration Studies, Washington, February 5, 1995. Joe Davidson, "Simpson to Drop Bill Provisions Curbing Job-Related Immigration," *Wall Street Journal*, March 7, 1996.

Chapter 18: Creating a Consensus

247. For a detailed discussion of the foresight concept and its application to the U.S. decision-making process, see Lindsey Grant, *Foresight and National Decisions: The Horseman and the Bureaucrat* (Lanham, MD: University Press of America, 1988).

248. See Joseph J. Brecher, "Population and the 'EIS'" (Teaneck, NJ: NPG, Inc., NPG FORUM series, May 1991) for a discussion of the judicial history of the EIS process as it applies to population.

249. "A Congenial Job for the Vice President," NPG Footnote series, 1993.

250. In December 1981, the Global Tomorrow Coalition, a broad coalition of most major U.S. environmental and population organizations and many futures study groups and interested individuals, unanimously adopted at its first annual meeting a position statement entitled *The Need to Improve National Foresight*. It called for the government to "establish in the Executive Office of the President an improved capacity to coordinate and analyze data . . . on the long-term interactions of trends in population, resources, and environment—and their relationship to social and economic development—and to provide information relevant to current policy decisions."

251. Lindsey Grant et al., *Elephants in the Volkswagen* (New York: W. H. Freeman and Co., 1992).

252. The graph is from Herman E. Daly, "The Steady-State Economy: Alternative to Growthmania" (Washington, D.C.: Population-Environment Balance, Monograph Series, April 1987). The quotation is from his "Allocation, distribution and scale: towards an economics that is efficient, just, and sustainable," from *Ecological Economics*, 6 (1992), p. 187.

253. Nathan Keyfitz, "The Biologist and the Economist: Is Dialogue Possible?" (Teaneck, NJ: NPG, Inc., the NPG FORUM series, June 1992).

254. Lindsey Grant et al., *Elephants in the Volkswagen* (New York: W. H. Freeman, 1992), p. 3.

GLOSSARY OF ACRONYMS

AFDC	Aid to Families with Dependent Children
AID	U.S. Agency for International Development
AIDS	Acquired Immune Deficiency Syndrome
AOSIS	Association of Small Island States
AP	Associated Press
APEC	Asia Pacific Economic Cooperation
BLM	Bureau of Land Management
CCC	Civilian Conservation Corps
CEO	chief executive officer
CEQ	Council on Environmental Quality
CFC	chlorofluorocarbon
CO2	carbon dioxide
CRP	Conservation Reserve Program
DDT	dichlorodiphenyltrichloroethane (pesticide)
DMZ	demilitarized zone
DNA	deoxyribonucleic acid
DOE	Department of Energy
EC	European Community
EIA	Energy Information Administration (DOE)
EIS	Environmental Impact Statement
EPA	Environmental Protection Agency
ERS	Economic Research Service
ESCWA	UN Economic and Social Commission for Western Asia
EU	European Union
FAIR	Federation for American Immigration Reform
FAO	Food and Agriculture Organization (UN)
FFC	fully fluorinated compound
FSU	former Soviet Union
G–7	"Group of Seven" Economic Summit countries (United States, Great Britain, Germany, Italy, Canada, France, and Japan)
GAO	Government Accounting Office
GATT	General Agreement on Tariffs and Trade
GDP	gross domestic product
GNP	gross national product
GPO	Government Printing Office
HCFC	hydrochlorofluorocarbon
HIV	human immunodeficiency virus
HUD	Housing and Urban Development
I=PCT	Impact equals Population times Consumption levels times Technology
ICPD	International Conference on Population and Development (UN)
ILO	International Labor Organization

INS	Immigration and Naturalization Service
IPCC	Intergovernmental Panel on Climate Change
IRCA	Immigration Reform and Control Act of 1986
ITO	International Trade Organization
LDC	less developed countries
MNC	multinational corporation
N2O	nitrous oxide
NAFTA	North American Free Trade Agreement
NAPAP	National Acid Precipitation Assessment Program
NAS	National Academy of Sciences
NATO	North Atlantic Treaty Organization
NEPA	National Environmental Policy Act of 1969
NOAA	National Oceanic and Atmospheric Administration
NOX	nitrate
NPCA	National Parks and Conservation Association
NPP	net primary productivity
NRDC	Natural Resources Defense Council
NSC	National Security Council
OECD	Organization for Economic Cooperation and Development
OTA	Office of Technology Assessment
PCB	polychlorinated biphenyl
PCSD	President's Council on Sustainable Development
PPFA	Planned Parenthood Federation of America
RNA	ribonucleic acid
SOX	sulfate
SSI	Supplemental Security Income
TFR	total fertility rate
TNC	transnational corporation
UN	United Nations
UNCED	United Nations Conference on the Environment and Development
UNDP	United Nations Development Programme
UNEP	United Nations Environment Programme
UNESCO	United Nations Educational, Scientific, and Cultural Organization
UNFPA	United Nations Fund for Population
UP	United Press
USDA	U.S. Department of Agriculture
USGS	U.S. Geological Survey
USITC	U.S. International Trade Commission
USLE	Universal Soil Loss Equation
USSR	Union of Soviet Socialist Republics
USWOC	U.S. Women of Color
UV	ultraviolet
VOA	Voice of America
WHO	World Health Organization
WRI	World Resources Institute
ZPG	Zero Population Growth

INDEX

A

Abortion, 12, 292–93n.93; consequences of to China, 122; and fertility, 252–53; issue of, 238–39; position on by environmental organizations, 252–53. *See also* Contraception

Accessible runoff, 29

Acid deposition, 59–61; and regional pollution, 57–61

Acid Deposition Control Program, 58

Acid emissions, 61

Acid fog, 59

Acid rain, 59, 60, 119

Acreage, expanding, 31. *See also* Farmland; Rangeland

Aerosols, 64, 66. *See also* Pollution

Affirmative Action, 160–61; meaning of, 234; as victim of demography, 161

Africa, 104–12; age distribution of, 106; and AIDS, 292n.89; population limits of, 105

Agendas, multiple, 237–50

Agricultural diversification, 41; productivity, and forests, loss of, 19; reward system, 165

Agriculture: by-products of commercial, 25–26; and energy transition, 51–52; genetic engineering, effect of on, 32; IPCC and effects on, 65–66; low-till, 89; "slash and burn," 71; U.S., 163–66; and weather, 43; yields in Africa, 107

AIDS: in Africa, 292n.89; and healthcare costs, 198–99; and population growth, impact of on, 11–12; worldwide cases of, 200

Algae, ocean, 42

Alienation, 245–46; and government, 247

Alliance, building, 254

Aluminum smelting, by-products of, 64

Ambivalence toward population movement, 257–58

Amphibians: decline of, 291n.69; experiments with, 79

Animal husbandry, aquaculture as, 41

Animals, food for rangeland, 72

Anthropogenic climate forcing, causes of, 68

Aquaculture, 38–39; as animal husbandry, 41

Aquatic systems, 65

Arable land. *See* Land

Architecture, solar, 93

Atmosphere: and automobiles, effects of on, 81–82; concentrations of chemicals in, 61; concentrations of greenhouse gases in, 63; and nitrogen, escape of into, 27

Atmospheric haze, 60

Automobiles: and atmosphere, 81–82; production of in China, 118

Avoidance, 254

B

Baselines, lack of, 16

Berlin Conference, 146–48

Biodiversity, 65; 96–98

Biologists, and creating consensus, 274–76

Biomass, 126; defined, 25; harvesting renewable, 70–72

Biomass-intensive scenario, 69

Biosphere: genetic engineering, effect of on, 32; and microbes, importance of, 96

Bituminous coal. *See* Coal

Borders, controlling, 212

Bosnia, 266

Breeder reactors, 53

Brundtland Commission, conclusions of, 123

C

Cairo conference, and feminists, 240–42. *See also* UN International Conference on Population and Development (ICPD)

Capitalism in China, 118

Carbon dioxide, 57; absorption of by forest "sinks," 71; costs of capturing, 70; effects of on climate forcing, 68; emissions, 63, 146, 147; and extinction, 290–91n.64

Carbon emissions, annual, 62

Cars. *See* Automobiles

Caution, politics of, 254–63

Center for Immigration Studies, estimates of annual totals, 159–60

Cereals: nitrogen-fixing, 32; production, 14, 25, 35–36. *See also* Grain

Change, technological, 179–80

Chemicals: man-made, 14; mutagens, dangers of, 15; releases, consequences of, 80–81

China, 112–22; statistics, official, 293n.94

Chlorofluorocarbons (CFCs), 79

Cities, disintegration of, 173–78

Classes, proportion of wealth between, in U.S., 175–78

Climate, 61–76; and Berlin Conference, 146–48; and coal, 50; effect of human activities on, 62–65; effects of fossil fuels on, 54; models, 64; stresses on, 64–65

Climate change, causes of, 62; and aerosol haze, 66; and desertification, 37

Climate warming, 75; effects on Earth's carrying capacity, 30, 60, 65–67, 74–75; engineering solutions to, 42; and fishery declines, 42; forcing, 68; global mean surface temperature, 63; proposed solution demands, 73–74. *See also* Global mean surface temperature; Global warming

Coal, 50; burning soft, 61; effects of on climate, 50; gasification plant, 59; lignite, 57; plans for, 69–70; U.S. resources of, 167; use of in China, 116, 118. *See also* Fossil fuels

Communism in China, 118

Condition of acreage, UN study on, 20

Consensus, creating a, 264–83

Conservation, 92–100; and energy transition, 51; and foresight, 93–94; and international policy, 277; limits of, 35–36; and petroleum supplies, 48; of water in U.S., 166

Conservationists, and creating consensus, 276–78

Constant fertility projections, 9, 10, 11; for Africa, 106–7. *See also* Fertility

Consumption, 146–49; consequences of rising, outlined, 76–77; levels, 8; optimization vs. maximization, 271–72; rate of oil, 8; over-, 249

Contraception, 12, 90. *See also* Abortion

Cooperation between rich and poor, 149

Corn yields, history of in U.S., 24

Crime rate, and unemployment, U.S., 175

Critical Trends Assessment Act, 270–71

Crops: diversity, 43; drought-tolerant, 30–31; salt-tolerant, 31; yields, and nitrogen, 23; yields, costs of high, 25–28

Crowding: costs of, 224–25; regulations, 222–23; social controls imposed by, 272

Cultural differences, 245

Cultural propensity; toward higher fertility, 209

D

Dams, 29; and siltation, 40

Data, official use of, 4

DDT, 27

Death rate from infectious diseases in U.S., 199

Democracy vs. plutocracy, 177

Demographic: changes, 160–61; cycles, 110; future, 160, 161–62; trends of Spain, Italy, Greece, Germany, 129

Demography: affirmative action as victim of, 161; consequences of, 267; "shifting shares" phenomenon, 209

Denial, 246–47

Dependency ratio, 107, 128

"Depensation," 288n.33

Desalinization for irrigation, 152. *See also* Salinization

Desertification, 18; and climate change, 37

Deserts, 65

Diseases: AIDS, 111–12; cycles of human, 98; death rate from infectious in U.S., 199; infectious, new and reemergent, 199; migration of, 132; reappearance of, 13; spread of infectious, 67

Disincentives, 240; and fertility, 207–8

Displacement: labor, 83–85; of peasants, 138; of peasants, 139–40

Diversification of energy resources in U.S., 168–69

Drought-tolerant crops, 30–31

E

Earth, limitations of, 20

Ecological systems, changing, 13

Economic aid, U.S. to Israel, 152

Economics, 230–31, 274–76; ecological, 274–75; relations, 86

Economy: of China, 117–19; rising prices as measure of, 249; as subsystem, 274

Ecosytem, development of, 55

Education: and fertility, 85, 208; as solution to unemployment, 190; universal primary, 241

Egypt, 152

Elderly: caring for, 248; medical/welfare costs of, 197; population of in U.S., 196–98

Emerging nations, 112–22, 141–44

Emigration. *See* Immigration, Migration

Emissions: of gas from greenhouses, 63; total ultimate, importance of, 75

Employment, and national health care program, 202

Energy, 167–70; and conservation, transition from, 51; conventional, 45–46; cost of solar/wind vs. fossil fuel, 74; diversification of base, 49; diversifying U.S. resources, 168–69; doubling of use of, 55; efficient use of, 124; and fertilizer production, 27; fossil, 31; generating from thorium, 289n.43; photovoltaic, 292n.81; renewable, 73–74, 87, 90; requirements, and agriculture, 51–52; scenarios, by IPCC, 69–70; sharing, 21; transition to renewables, 45–55, 126; transition, and food, 51–52; unconventional sources of, developing, 48; uranium as transitional source of, 53; use in Japan, 170; use per capita, 45–46

Entropy, 275

Environment, 144–50; cleanup of, 58–59; effects of fossil fuels on, 54; and industry, 117–19; problems, international ramifications of, 145; stresses on, effects of, 64–65; and technology as source of problems, 78–79; and trade, 139

Environmental organizations, and various policies, endorsement of, 251–54

Environmental Protection Agency (EPA), recommendations of, 271

Epidemics: and migration, 200; preventing, importance of, 199–200; reappearance of, 13

Equality, women's, 242

Erosion, 164–65; in China, 114; effects of on farmland, 18

Ethnicity, 160–61

Europe, fertility rates of Northern and Western,

129
European Union, food surplus of, 126
Eutrophication, 28. *See also* Groundwater; Oxygen, depletion of
Extinction, 290–91n.64

F
Family planning: in China, 117; and foreign aid, 214–15; position on by environmental organizations, 252–53; and women's rights, 239
Famine: and nitrogen, 23; potato, 21–22
Farming: agricultural reward system, 165; sustainable, 165
Farmland: conversion of forests/range to, 18; conversion of to rangeland in U.S., 164; and erosion, 114; expanding, 31; and forests, competition with, 19–20; and industrial use, competition with, 20; loss of, effects of, 18. *See also* Land; Rangeland
Federal farm support program, 165
Feminism, 253; and Cairo conference, 240–42
Fertile land. *See* Land
Fertility: and abortion, 252–53; average world, 9; below replacement level, countries with, 125; and contraception/abortion, 12; declines in third world, 10; decline, causes of, 11; differential, 144; discouraging high, 208; and education, 208; in industrialized countries, 12; and labor market, effect of on, 85; levels of, summary by region, 16*fig.*; levels of for different countries, 10; levels of for various countries, 10; managing, 35; and national health plan, impact of on, 201–2; patterns of and industrialization, 11; rewarding lower, 191; statistical conventions about, 11; "stop at two," 204–9; total rate of, 8–9; and U.S., 157–60; as variable influencing population growth, 204; and women's rights, 239. *See also* Constant fertility projections
Fertility rates: of Germany, 128; of Northern and Western Europe, 129
Fertilization of sea with iron, 42
Fertilizers: chemical, 25; heavy use of, 72; intensive use of by emergent countries, 125; quintupling of use of, 71; synthesizing nitrogen, 96; synthetic, 26, 33; and water quality, 28; world use of, 25
FFCs (fully fluorinated compounds), 64
Fiber and food, competition with, 70–72
Fish: annual catch of wild, 39; populations, 40–41; stocks, depletion of, 39
Fisheries, 38–39; government subsidization of, 40
Fission, 52–53. *See also* Nuclear energy
Fog, acid, 59
Food: basic, 21–28; competition for, 123; consumption, estimated worldwide, 38; distribution of, 35; emergency world needs, 108; and energy transition, 51–52; and fiber, competition with, 70–72; future needs, and technology, 31–33; as limiting factor, 17; production, 14, 24; production of in Africa, 107; production, and chemical fertilizers, 25; prospects for, 17–44; supply problems, 24; world security, 136; yield, and nitrogen, 23
Forces: generating basic world problems, 6; tectonic, and population growth, 7
Foreign aid, 213–16; family planning and, 214–15; to Israel from U.S., 151–53; problems resulting from, 110
Foreign policy, 265
Foresight, 267–68
Forests: as "sinks" for carbon absorption, 71; conversion of to farmland, 18; and farmland, competition with, 19–20; hardwood, and urbanization, 164; Intergovernmental Panel on Climate Change (IPCC) and effects on, 65; threat to systems, 60; tropical, loss of, 19
Fossil energy, 31
Fossil fuels, 57–58; climatic/environmental effects of on, 54; cost of vs. solar or wind energy, 74; issue of to developing countries, 124; rate of burning of, 75; tax, President Clinton's, 51; tax, in U.S., objectives of, 168–70; transition away from, 50–51; transition from renewables, 45–55. *See also* Coal; Gas; Petroleum
Free trade: abandoning, 188–89; Alliance, as threat to national environmental legislation, 139; and American jobs, 187; dangers of drive toward, 143; and job flight, 180–81; and MNCs, 180–81; and the poor, 139–40; proponents of, 141. *See also* North American Free Trade Agreement (NAFTA)
Free-trading system, 142
Freon. *See* Chlorofluorocarbons (CFCs)
Fuel: cell engine, 93; methanol vs. hydrogen fuel cell, 82; synthetic, 20
Fully fluorinated compounds (FFCs), 64
Fungus, potato, 22, 26. *See also* Potato famine
Fusion, 54–55. *See also* Nuclear energy

G
Gas, 49–50; greenhouse, emission levels from, 63, 73; U.S. resources of, 167. *See also* Fossil fuels
Gasification plant, 59
Genetic engineering, 33; and biosphere, effect of on, 32; of crops, 20; use of in agriculture, 32
Genetic manipulation, 97
Geographical conventions, 285–86
Germany, 128
Globalization, 137
Global mean surface temperature, 63. *See also*

Climate warming; Global warming

Global sea level, 63

Global warming, 30, 167; avoiding, 82. *See also* Climate warming; Sea level

Government relevancy, 265–67

Grain: consumption of, 35; importation, 114–15; prices, 17; production, 37; production in China, 115; world trade, 135–37; yields, U.S., 163. *See also* Cereals

Grazing: modern practices of, 37; rights, 36

Greenhouse effect, 61, 123; gas emissions, 63, 73, 147; hastening of, 64

Groundwater: contamination and wetlands, effects of on, 28–29; contamination of, 28; nitrates in Dutch, 126. *See also* Eutrophication; Water

"Group of Seven" (G-7) industrial nations, 130

Growth, opposition to, 220–26

Growth/collapse cycle, 38

H

Haber-Bosch process, 23, 31

Habitat, human, and health, 66–67

Health care: and AIDS, 198–99; elderly, costs, 248; medical/welfare costs, 197; national program in U.S., 201–2; plagues, cost of, 199–200; program, and employment, 202; reform, 195–202; universal, 241

Herbicides, 27

"High demand variant" scenario, 69

Hostility toward population movement, 257–58

Human density projections, 8

Human needs, 4

Human population, level of sustainable, 74–75

Human scale, 223–24

Hydrochlorofluorocarbons (HCFCs), 79. *See also* Pollution

Hydrogen fuel cell, 82

Hydroponics, 32

I

Ice caps, effects of complete melting of, 67–68

Identification: establishing, 211; system of, 200

Immigrants: contributions of, 247–48; illegal, 134, 159–60

Immigration, 127; and environmental damage, 252; estimates of annual totals, 159–60; of hungry, 178–79; and jobs, 184–87; and labor, 243–44; legal, 160; and menial jobs, 246; policy, 184–85, 162, 268–69; position of Roman Catholic church on, 246; reform, 209–12; Roper poll on, 261; U.S. and, 157–60; as threat to U.S. wages, 187; and welfare, 194. *See also* Migration

Incentives, 240; and fertility, 207–8

India, total fertility rate of, 123

Individualists; vs. societalism, 226–33; viewpoint of, 227

Individuals, rights of, 228–29

Industrial: definition of, 125; countries, 12, 140–41; world, the, 125–31

Industrialization: process, and fertility patterns, 11; results of in developing countries, 123; success of in West, 146

Industry: and Chinese peasantry, 116–17; and environment, 117–19; and farmland, 20; and forests, competition with, 20; and waste, 94–95. *See also* Labor; Trade

Innumeracy, 244–45

Instabilities between world systems, 6

Interdependence, 86–88

Intergovernmental Panel on Climate Change (IPCC): on consequences of climate warming/rising sea levels, 65–67; energy scenarios, 69–70; and global warming, counteracting, 148; handling of population growth as independent variable, 72; projection of human effect on, 62–65

Internal combustion engine, 93–94

International policy, and conservation, 277

Ionizing radiation, 53

Irish potato famine, 21–22

Irrigation, 29, 30, 114; desalinization for, 152; expense of, 30

Israel: fertility rate of, 151; U.S. foreign aid to, 151–53; and imports, dependence on, 151

J

Japan: energy use in, 170; and food import, 136

Job balance, between rich and poor, 175–78

Joblessness. *See* Unemployment

Jobs: and immigration, 184–87; menial, and immigration, 246; and welfare, 189–94. *See also* Employment; Labor

Juggernaut, metaphor of, 2

K

Killer fogs, 56–57

Krill, 42

L

Labor: displacement, 83–85; in Europe, 129; low cost competition, 141–44; MNCs and displacement of peasants, 138. *See also* Employment; Industry; Jobs; Unemployment

Labor force, role of women in, 181

Labor policy, and trade policy, 187

Land: amount of per capita, 17–18; arable, assumptions about, 33; arable, per capita in Africa, 107; and food production, 17–21; revival of, 18–19; U.S. use of, 163–72. *See also* Farmland; Rangeland

Leadership, 251–63

Legal immigration, 160

Legislation and population, 261–63

Life expectancy, 10, 11

Lignite. *See* Coal
Logistic curve, 23
Losses from pests/weeds, 26–27
Low wage competition, 141–44

M

Majority, current definition of U.S., 161
Malthus, Thomas, and population debate, 34–35
Meat: 36–38; as cereal production derivative, 38; production, 37
Methane, effects of on climate forcing, 68
Methyl bromide, 80
Microbes, importance of, 96
Microorganisms, 60–61
Middle class, and proportion of wealth, 177
Migration, 133–35; border control and, 212; generation of by China, 121; of hungry, 178–79; interagency group on, 268–69; long distance, 134; of plagues and diseases, 132; UN International Conference on Population and Development (ICPD) on, 260; U.S., 159–60; as variable influencing population growth, 204; as wild card, 11. *See also* Immigration
Mineralization of water, 29
Monoculture(s), 22, 26
Mortality, 10; rising rate, effect of, 9; statistical conventions about, 11
Multinational corporations (MNCs), 137–38; free trade and, 140, 180–81
Mutation: pathogenic, 98; of pests, 26–27

N

National Academy of Sciences (NAS), chemicals survey results, 14–15
National Commission on Population Growth and the American Future, recommendations of, 255–56
National Environmental Policy Act of 1969, purposes of, 269–70
National health care program. *See* Health care
National population policy, 256–57
Natural gas as feedstock and energy source, 27
Natural gas-intensive scenario, 69
Neurological toxicants, 15
Nitrogen, 22–23; and atmosphere, escape of into, 27; sources of, 32; synthesizing fertilizer, 96
Nitrous oxide: and atmosphere, 27; effects of on climate forcing, 68
Nonindigenous populations, 238
Nonlinearity, 99
North American Free Trade Agreement (NAFTA), 138, 143; and unemployment, 144. *See also* Free trade
Nuclear energy, 52–55; dangers of, 54; fission/fusion, 52–55
Nuclear fission. *See* Fission
Nuclear-intensive scenario by IPCC, 69

Nuclear waste, storage methods for, 53

O

Ocean algae, 42
Ocean commons, regulation of, 40–41
Ocean resources, 41–42
Ocean temperatures, 67
Oil, U.S. resources of, 167. *See also* Petroleum
One-child policy of China, 112, 120
"One World," 242–46
Optimism, 280–83
Organic compounds, volatile, decline of, 58
Organic farming, 25–26
Orphanages, 292–93n.93
Overconsumption, 249
Oxygen, depletion of, 57. *See also* Eutrophication
Ozone: levels, 60; stratospheric, 64, 79, 80; and ultraviolet radiation, 79–80

P

Paradigms, 217–24
Passivity, 263
Pastoral systems, 37
Patriotism, 233–36
Peasant labor and reforestation, 19
Perpetual growth model, alternative to, 273–74
Pesticides: effects of, 26–27; resistance to, 43
Pestilence, 132–33; benefits of, 112
Pests, losses from all, 26
Petroleum, 46–49; concentrations of in Persian Gulf, 49; and conservation, 48; current consumption levels of, 47; export of, 115–16; and waste, creation of, 81. *See also* Fossil fuels; Oil
Phosphates, world resources for, 72
Photovoltaic energy production, 292n.81
Physical systems, changing, 6
Phytophthora infestans, 22. *See also* Irish potato famine
Plagues: and health care costs, 199–200; migration of, 132
Plankton, 40, 42, 79
Plant growth, limiting factors in, 23
Plutocracy vs. democracy, 177
Plutonium, 53
Policy: central question about, 11; population, 184–202
Politics of caution, 254–63
Pollutants and synergy, 59–60
Pollution: by automobiles, 81–82; avoiding, 79–80; in China, 118; and coal, 50; "competitive right to pollute," 148–49; cooling effect of aerosol, 64; "credits," 147; of European rivers, 126; and lignite, 57; local, and killer fogs, 56–57; regional, and acid deposition, 57–61; "right to pollute," 146–49; smog, 56–57; sources of, 58; sulfur dioxide, 61; tax in U.S., objectives of, 168–70
Population: curve for 20th century, 13; decline

in Europe, 11; destabilization of poorer countries,' 134–35; growth, and future of U.S., 157–63; growth, consequences of, outlined, 76–77; growth, ripple effect of, 132–35; growth, this century, 7; history, 16; legislation, 261–63; levels, 8; limits, regions reaching, 105; momentum, 107; movement, hostility/ambivalence toward, 257–58; nonindigenous, 238; policy, 184–202; policy, U.S. national, 256–57; program assistance to Egypt, 152; programs in China, 112–14; projections, 10; solutions, 203–17; stabilization, and third world leaders, 150; taboo, 237–50; third world growth, United States policy on, 260; variables influencing U.S. growth, 204; working age in U.S., 179

Postwar crisis of industrial world, 129–31

Potato famine, 21–22

Precision vs. trend, 4

Pregnancy, reasons for teen, 207

President's Council on Sustainable Development, 258–59

Prices as measure of economy, 249

Processes for converting nitrogen, 23

Protectionism, 130

Protein, sources of, 38, 39

R

Race, 160–61

Racism, 249–50

Radiation: ionizing, 53; solar, 64; ultraviolet, 40; ultraviolet, and ozone, 79–80

Rain, acid, 59, 60

Rangeland, 36; animals, food for, 72; conversion of farmland to in U.S., 164; conversion of to farmland, 18; deterioration of, 37. *See also* Farmland

Recycling: argument for, 171; and nutrients, 95; water, 33

Reform, immigration, 210–11

Regional effects of climate warming/rising sea levels, 66. *See also* Climate warming; Global warming

Regional pollution, 57–61

Relativism, 245

Renewable energy, 90; cost of, 73–74; transition from fossil fuels to, 45–55

Renewables, 70–72

Replacement level fertility, 10

Republican alternative, 260–61

Resources: competition for limited, 272; diminishing base of, 181–82; energy, 167; recoverable, 289n.38

Rights: of individuals, 228–29; of women, 239

Roman Catholic church position on immigration, 209, 246

Roper poll on immigration, 261

S

Salinization, 18, 30. *See also* Desalinization

Salt-tolerant crops, developing, 31

Sanitation, 13

Sea ice, effects of retreat of, 67

Sea level; global rise of, 63; rising, effects of, 65–67. *See also* Global warming

Seaweed, 41

Second Assessment, magnitude of, 74

Sewage: in China, 118; and human population, connection between, 95

"Shifting shares" phenomenon, 209

Sixteenth Amendment, dismembering the, 230

"Slash and burn" agriculture, 71

Social Contract, 228, 235–36

Social justice, 237

Social obligations vs. individual rights, 228–29

Social systems, changing, 6

Societalists: vs. individualists, 226–33; viewpoint of, 227

Society, 173–83

Socioeconomic and environmental systems, instabilities between, 6

Soil: microorganisms, 60–61; organic materials in, loss of, 25–26; worsening conditions of, worldwide, 18

Solar: construction, 93; energy, cost of vs. fossil fuel cost, 74

Specter/Abraham amendment, 262

Stability, 126–29

Standard of living, 126

Statistical conventions about fertility/mortality, 11

Statistics, trusting, 3–4

Steady state economy, 274

Stock carryover, worldwide, 17

Stratosphere, ozone depletion of, 64

Sulfur dioxide, 57, 61

Surveys, list of, 284–85

Sustainability, 170–71, 275

Sustainable farming, 165

Synergy, 6, 98–99; and pollutants, 59–60

Synthetics: fertilizers, 23; fuels, 20; nitrogen, 23

T

Taboos, population, 237–50

Technological revolution, narrowing effects, 87–88

Technology, 78–91; benefits of changes in, 139–40; as destroyer, 88–89; and environmental problems, 78–79; and food needs, future, 31–33; impact of on unemployment, 179–80; and labor displacement, 83–85; and unemployment, 82–85; uses of, 89–91; and waste, 80–82

Tectonics defined, 6

Temperature, global mean surface, 63. *See also*

Climate warming; Global warming
"Terrestrial net primary productivity" (NPP), 21
Third world population growth, U.S. policy on, 260
Thorium, generating energy from, 289n.43
Total fertility rate (TFR), 10, 204; in Africa, 8–9; in India, 9; of India, 123
Total ultimate emissions, importance of, 75
Toxicity of chemicals, 15
Toxins, "biochemical remediaton" of, 90
Trade, 140–41; barriers, 136; deficit, U.S. and China's, 119; policy, and labor policy, 187; unrestrained free, 140–41. *See also* Industry
Transnational corporations (TNCs). *See* Multinational corporations (MNCs)
Trees, damage to in Europe, 61
Trends: in China, 122; destructive, 182–83; in Europe, 11; and job balance between rich and poor, 175–78; vs. precision, 4
Tropical forests. *See* Forests
Two-child family, 204–9

U

Ultraviolet (UV) radiation, 40; level of, 79; and ozone, 79–80
UN. *See* United Nations (UN)
Undocumented aliens. *See* Immigrants, illegal
Unemployment: benefits of, 190–91; and Chinese peasantry, 116–17; education as solution to, 190; European, 127; and fishery policy, 40; impact of technology on, 179–80; and migration, 133; sources of, 178–82; and technology, 82–85. *See also* Labor
Unemployment statistics, U.S., 174
United Nations (UN): Environment Program (UNEP) study on condition of acreage, 20; projections, 8–9, 12; UN International Conference on Population and Development (ICPD), 8, 240–42, 259–60; world "constant fertility" projection, 8
United States: and abortion, consequences of, 122; demographic history of, 157; and fertility, 157–60; future of and population growth, 157–63; government and UN International Conference on Population and Development (ICPD), 259–60; and migration, 157–60; national population policy on various issues, 256–57; policy on third world population growth, 260; trade deficits, 119
Universal health care, 241
Universalism, 243–44
Universal primary education, 241
Uranium, 81; as energy source, transitional, 53
Urban growth projections by UN, 12
Urbanization, 12–13; and hardwood forests, 164

V

Variables affecting demographic future, 160
Vatican, position on immigration, 246
Violence: and declining resources, 40; and labor displacement, 83; and unemployment, U.S., 175
Volatile organic compounds, decline of, 58

W

Wages, 140–41. *See also* Employment
Waste: industry and energy, 94–95; nuclear, storage methods for, 53; technology and, 80–8
Water, 28–33, 150–53; availability, varying, 29; in California, use of, 288n.25; conservation, U.S., 166; desalinization costs, 152; distribution, 295n.149; U.S. financing of projects, 295n.147; ionizing for fuel, 82; as limit, 29–31; as major problem for Israel, 151–53; pollution, 57; quality, and fertilizers, 28; "ranching," 31; resources, 65; recycling, 33; scarcity, 30; shortage of in U.S. Southwest, 166; tables, 29, 171. *See also* Eutrophication; Groundwater
Wealth, pursuit of, 145–46
Weather and agriculture, 43
Weeds, 27, 32
Welfare: costs of elderly, 197; and immigration, 194; and jobs, 189–94; outline of policy for, 192–94; restructuring, 191; society of Europe, 130
Wetlands: coastal, destruction of, 39–40; and groundwater contamination, effects of on, 28–29
Wind energy, cost of vs. fossil fuel cost, 74
Women: equality of, 242; rights of, and family planning, 239; role of in labor force, 181
World "constant fertility" projection by UN, 8
World environmental and socioeconomic systems, instabilities between, 6
World grain trade, 135–37
Worldwide stock carryover, 17

Y, Z

Yields, dependence on high, 28
Zero growth and sustainability, 171